AF520201

Molecular Techniques in Crop Improvement

THE AUTHOR

Dr. Phundan Singh [b. 1946] hails from a reputed agricultural family of Western Uttar Pradesh [Village-Sakauti, Tehsil- Mawana, District- Meerut]. He graduated in Agriculture from Agra University, Agra and obtained his Master and Doctorate degrees from Kanpur University, Kanpur. He has throughout a brilliant academic record. His area of specialization is Plant Breeding and Genetics. Dr. P. Singh has total experience of more than 40 years in the field of Genetics and Plant Breeding. He has served the Central Institute for Cotton Research, Nagpur for about 32 years in various capacities such as Scientist, Senior Scientist, Principal Scientist, Head of Division and Director. He has 300 publications to his credit. He has participated and presented 80 research papers in various National and International Seminars, Symposia and Conferences He has authored more than 90 books on Genetics, Plant Breeding, Seed Technology, Biotechnology, *etc.* mostly published by Kalyani Publishers, New Delhi. He has also authored 18 technical bulletins on various aspects of cotton and contributed 14 chapters in various books. Dr. P. Singh is a member of several Scientific Societies, referee of different journals, Expert member of Selection Committees and M. Sc. and Ph. D. Examiner in different Agricultural Universities. The Institute was recognized as the referral Laboratory for detection of Bt. gene and received the best Annual Report Award when he was the Director. He has visited several countries such as Russia, Belarus, Ukraine, Crimea, Uzbekistan, Canada, USA, England, Austria and Germany.

Molecular Techniques in Crop Improvement

Phundan Singh

Former Director
ICAR-Central Institute for Cotton Research,
Nagpur – 440 010 Maharashtra, India

2022

Daya Publishing House®
A Division of
Astral International Pvt. Ltd.
New Delhi – 110 002

© 2022 AUTHOR
ISBN: 9789354616310 (Int. Edition)

Publisher's Note:

Every possible effort has been made to ensure that the information contained in this book is accurate at the time of going to press, and the publisher and author cannot accept responsibility for any errors or omissions, however caused. No responsibility for loss or damage occasioned to any person acting, or refraining from action, as a result of the material in this publication can be accepted by the editor, the publisher or the author. The Publisher is not associated with any product or vendor mentioned in the book. The contents of this work are intended to further general scientific research, understanding and discussion only. Readers should consult with a specialist where appropriate.

Every effort has been made to trace the owners of copyright material used in this book, if any. The author and the publisher will be grateful for any omission brought to their notice for acknowledgement in the future editions of the book.

All Rights reserved under International Copyright Conventions. No part of this publication may be reproduced, stored in a retrieval system, or transmitted in any form or by any means, electronic, mechanical, photocopying, recording or otherwise without the prior written consent of the publisher and the copyright owner.

Published by : **Daya Publishing House®**
A Division of
Astral International Pvt. Ltd.
– ISO 9001:2015 Certified Company –
4736/23, Ansari Road, Darya Ganj,
New Delhi-110 002
Ph. 011-43549197, 8130496929
E-mail: info@astralint.com
Website: www.astralint.com

— Dedicated to —

Mrs. Jaswanti Singh [Wife]

Engineer Rajeev Singh[Son]

Ms. Pavitra Choudhury [Daughter]

Miss Dr. Rishika Singh [Granddaughter]

Preface

Molecular techniques have wide practical applications in genetics, plant breeding, Biotechnology, Seed technology, *etc.* However, information pertaining to molecular techniques is not available from any single source. The prime objective of writing this book is to fulfill this long felt need of Indian students. Main features of this book are as follows:

1. This book provides comprehensive information related to molecular techniques in one compact volume of 20 chapters. Efforts have been made to provide up to date information even then some valuable information has been left out.
2. Each chapter has been presented step by step and in a simple language for easy grasping of students. Chapter wise Glossary is at the end of book for ready reference of readers. Some useful references are also given at the end.

The information contained in this book has been gathered from internet and various published sources. The cooperation extended by my wife Mrs. Jaswanti Singh during preparation of this revised manuscript is highly appreciable. Hope, this book would be useful to the students, teachers and researchers engaged in the field of Plant Biotechnology, Genetics and Plant Breeding. Constructive suggestions of students and teachers are invited for further improvement of this book.

Phundan Singh

Contents

Chapter 1

Introduction to Molecular Techniques

INTRODUCTION

Common methods which are used in molecular biology, biochemistry, genetics and biophysics for manipulation and analysis of DNA, RNA, protein, and lipid are referred to as molecular techniques. There are several molecular techniques which are widely used in molecular genetics. Important techniques of molecular genetics include (1) gel electrophoresis, (2) polymerase chain reaction [PCR], (3) gene cloning, (4) gene sequencing, (5) southern blot technique, (6) northern blot technique, (7) western blot technique, (8) gene silencing, (9) DNA fingerprinting, (10) DNA microarrays, (11) fluorescence *in situ* hybridization, (12) gene mapping, (13) allele specific oligonucleotides, *etc.* A brief description of all these techniques is presented as follows:

1. Gel Electrophoresis

Gel electrophoresis is one of the principal tools of molecular biology. The basic principle is that DNA, RNA, and proteins can all be separated by means of an electric field. In agarose gel electrophoresis, DNA and RNA can be separated on the basis of size by running the DNA through an agarose gel. Proteins can be separated on the basis of size by using an SDS-PAGE gel, or on the basis of size and their electric charge by using what is known as a 2D gel electrophoresis. There are many forms of electrophoresis in use but all operate by putting a solution of the molecules to be separated into an electric field. In all electrophoresis experiments electrodes are positioned at opposite ends of the separation matrix which may be in the form of a tube or of a thick or thin slab. In the most common forms of electrophoresis two type of matrices, *viz..* agarose and polyacrylamide are used. There are many different commercial and homemade apparatuses in use. For unimportant technical reasons, acrylamide gels are usually held vertically and the samples are loaded in wells formed in the top edge whereas agarose gels are usually held horizontally and the samples are inserted into wells in their top face.

TABLE 1-1: Sieving Properties of Agarose and Polyacrylamide Gels

Sl.No.	Type of Matrix	Optimum Size Range of Molecules to be Separated
1	Agarose gel	0.5-20 kb
2	Polyacylamide gel	DNA: 10-1000 kb Protein: 0-100 kb

2. Polymerase Chain Reaction

The **polymerase chain reaction** (**PCR**) is a technique widely used in molecular biology. This technique was developed in 1983 by Kary Mullis for which she was awarded Nobel prize in1993. It derives its name from one of its key components, a DNA polymerase used to amplify a piece of DNA by *in vitro* enzymatic replication. As PCR progresses, the DNA thus generated is itself used as template for replication. This sets in motion a chain reaction in which the DNA template is exponentially amplified. With PCR it is possible to amplify a single or few copies of a piece of DNA across several orders of magnitude, generating millions or more copies of the DNA piece. PCR can be extensively modified to perform a wide array of genetic manipulations.

Almost all PCR applications employ a heat-stable DNA polymerase, such as Taq polymerase, an enzyme originally isolated from the bacterium *Thermus aquaticus*. This DNA polymerase enzymatically assembles a new DNA strand from DNA building blocks, the nucleotides, using single-stranded DNA as template and DNA oligonucleotides (also called DNA primers) required for initiation of DNA synthesis. The vast majority of PCR methods use thermal cycling, *i.e.*, alternately heating and cooling the PCR sample to a defined series of temperature steps. These thermal cycling steps are necessary to physically separate the strands (at high temperatures) in a DNA double helix (DNA melting) used as template during DNA synthesis (at lower temperatures) by the DNA polymerase to selectively amplify the target DNA. The selectivity of PCR results from the use of primers that are complementary to the DNA region targeted for amplification under specific thermal cycling conditions.

Applications

PCR is now a common and often indispensable technique used in medical and biological research labs for a variety of applications. These include DNA cloning for sequencing, DNA-based phylogeny, or functional analysis of genes; the diagnosis of hereditary diseases; the identification of genetic fingerprints (used in forensic sciences and paternity testing); and the detection and diagnosis of infectious diseases. In 1993 Mullis won the Nobel Prize in Chemistry for his work on PCR.

3. Gene Cloning

The process by which multiple copies of a specific gene (DNA) can be made after isolation of such gene or DNA is known as gene cloning. In other words, gene cloning is the process of making several copies of a gene. Gene cloning is known by a variety of names such as DNA cloning, molecular cloning and recombinant DNA technology. In gene cloning, a segment of DNA representing a gene is inserted in to a cloning vector where it replicates further. The c-DNA clones may be made for various purposes such as (i) expression studies (ii) gene structure and function studies, and (iii) gene tagging.

4. DNA Sequencing

The process of determining the order of the nucleotide bases along a DNA strand is called sequencing. In 1977, two separate methods for sequencing DNA *viz..* (i) chain termination method, and (ii) chemical degradation method were developed. In the beginning both methods were equally popular. However, today the chain termination method is more commonly used. This method is based on the principle that single-stranded DNA molecules that differ in length by just a single nucleotide can be separated from one another using polyacrylamide gel electrophoresis.

The DNA to be sequenced, called the template DNA, is first prepared as a single-stranded DNA. Next, a short oligonucleotide is annealed, or joined, to the same position on each template strand. The oligonucleotide acts as a primer for the synthesis of a new DNA strand that will be complementary to the template DNA. This technique requires that four nucleotide-specific reactions—one each for G, A, C, and T—be performed on four identical samples of DNA. The four sequencing reactions require the addition of five components, *viz..* (i) a DNA template, (ii) a primer tagged with a mildly radioactive molecule or a light-emitting chemical, (iii) DNA polymerase, an enzyme that drives the synthesis of DNA, (iv) four deoxynucleotides (G, A, C, and T), and (v) one dideoxynucleotide, either ddG, ddA, ddC, or ddT

After the first deoxynucleotide is added to the growing complementary sequence, DNA polymerase moves along the template and continues to add base after base. The strand synthesis reaction continues until a dideoxynucleotide is added, blocking further elongation. This is because dideoxynucleotides are missing a special group of molecules, called a 3′-hydroxyl group, needed to form a connection with the next nucleotide. Only a small amount of a dideoxynucleotide is added to each reaction, allowing different reactions to proceed for various lengths of time until by chance, DNA polymerase inserts a dideoxynucleotide, terminating the reaction. Therefore, the result is a set of new chains, all of different lengths.

To read the newly generated sequence, the four reactions are run side-by-side on a polyacrylamide sequencing gel. The family of molecules generated in the presence of ddATP is loaded into one lane of the gel, and the other three families, generated with ddCTP, ddGTP, and ddTTP, are loaded into three adjacent lanes. After electrophoresis, the DNA sequence can be read directly from the positions of the bands in the gel. The Sanger method is also known as the "dideoxy chain termination method".

5. Southern Blotting

This technique was developed by Edwin Southern in 1975. It is a method which is routinely used in molecular biology to detect the presence of a specific DNA sequence in a DNA sample. DNA samples after digestion with restriction enzymes are separated by gel electrophoresis and then transferred to a membrane by blotting via capillary action. The membrane is then exposed to a labeled DNA probe that has a complement base sequence to the sequence on the DNA of interest. Most of the original protocols used radioactive Label. However non-radioactive alternatives are now available. Southern blotting is less commonly used in laboratory science due to the capacity of other techniques, such as PCR, to detect specific DNA sequences from DNA samples. These blots are still used for some purposes such as detection of transgene and their copy number in transgenic organisms.

6. Northern Blotting

This technique was developed in 1977 by James Alwine, David Kemp, and George Stark at Stanford University. The northern blot is a technique used in molecular biology research to study the expression patterns of a specific type of RNA molecule among a set of different samples of RNA. It takes its name from the similarity of the procedure to the Southern blot procedure, named for biologist Edwin Southern, used to study DNA, with the key difference that, in the northern blot, RNA, rather than DNA, is the substance being analyzed by electrophoresis and detection with a hybridization probe. The gels may be run on either agarose or denaturing polyacrylamide gels depending on the size of the RNA to be detected. In this process RNA is separated based on size and is then transferred to a membrane that is then probed with a labeled complement of a sequence of interest. The results may be visualized through a variety of ways depending on the label used; however, most result in the revelation of bands representing the sizes of the RNA detected in sample. The intensity of these bands is related to the amount of the target RNA in the samples analyzed

A notable difference in the procedure in case of agarose gels, (as compared with the Southern blot) is the addition of formaldehyde which acts as a denaturant. For smaller fragments denaturing polyacrylamide urea gels are employed. As in the Southern blot, the hybridization probe may be made from DNA or RNA. This procedure is commonly used to determine time and conditions of expression of certain genes in living tissues.

7. Western Blotting

This technique was developed by Towbin and coworkers in 1979 which has originated from the laboratory of George Stark at Stanford. The western blot is a method of detecting specific proteins in a given sample of tissue homogenate or extract. The method originated from the laboratory of George Stark at Stanford. The name western blot was given to the technique by W. Neal Burnette and is a play on the name Southern blot, a technique for DNA detection developed earlier by Edwin Southern. Detection of RNA is termed northern blotting.

It uses gel electrophoresis to separate native or denatured proteins by the length of the polypeptide (denaturing conditions) or by the 3-D structure of the protein (native/non-denaturing conditions). In this technique, proteins are first separated by size, in a thin gel sandwiched between two glass plates in a technique known as SDS-PAGE (sodium dodecyl sulfate polyacrylamide gel electrophoresis). The proteins in the gel are then transferred to a PVDF, nitrocellulose, nylon or other support membrane. This membrane can then be probed with solutions of antibodies. Antibodies that specifically bind to the protein of interest can then be visualized by a variety of techniques, including colored products, chemiluminescence, or autoradiography. Often, the antibodies are labeled with an enzymes. When a chemiluminescent substrate is exposed to the enzyme it allows detection. Antibodies to most proteins can be created by injecting small amounts of the protein into an animal such as a mouse, rabbit, sheep, or donkey (polyclonal antibodies)or produced in cell culture (monoclonal antibodies). These antibodies can be used for a variety of analytical and preparative techniques.

This technique not only allows detection but also quantitative analysis.This method is used in the fields of molecular biology, biochemistry, immuno-genetics and medical sciences. Other related techniques include using antibodies to detect proteins in tissues and cells by immuno-staining and enzyme-linked immune-sorbent assay (ELISA). In medical science, western blot technique is used as (i) confirmatory test for HIV, (ii) definite test for Bovine spongiform encephalopathy (BSE, commonly referred to as 'mad cow disease'), and (iii) definitive test for Some forms of Lyme disease testing

8. Gene Silencing

Gene silencing is a general term describing epigenetic processes of gene regulation. The term gene silencing is generally used to describe the "switching off" of a gene by a mechanism other than genetic modification. That is, a gene which would be expressed (turned on) under normal circumstances is switched off by machinery in the cell.

Genes are regulated at either the transcriptional or post-transcriptional level. Transcriptional gene silencing is the result of histone modifications, creating an environment of heterochromatin around a gene that makes it inaccessible to transcriptional machinery (RNA polymerase, transcription factors, *etc.*). Post-transcriptional gene silencing is the result of mRNA of a particular gene being destroyed. The destruction of the mRNA prevents translation to form an active gene product (in most cases, a protein). A common mechanism of post-transcriptional gene silencing is RNAi.

Both transcriptional and post-transcriptional gene silencing are used to regulate endogenous genes. Mechanisms of gene silencing also protect the organism's genome from transposons and viruses. Gene silencing thus may be part of an ancient immune system protecting from such infectious DNA elements. Genes may be silenced by DNA methylation during meiosis, as in the filamentous fungus *Neurospora crassa.*

9. DNA Fingerprinting

DNA fingerprinting (also called genetic fingerprinting, DNA testing, DNA typing, or DNA profiling) is a technique used to distinguish between individuals of the same species using only samples of their DNA. Although two individuals will have the vast majority of their DNA sequence in common, DNA profiling exploits highly variable repeat sequences called VNTRs. These loci are variable enough that two unrelated humans are unlikely to have the same alleles. The technique was first reported in 1984 by Dr. Alec Jeffreys at the University of Leicester, and is now the basis of several national DNA identification databases.

Since the late 1950s and early 1960s, molecular biologists have learned to characterize, isolate, and manipulate the molecular components of cells and organisms. These components include DNA, the repository of genetic information; RNA, a close relative of DNA whose functions range from serving as a temporary working copy of DNA to actual structural and enzymatic functions as well as a functional and structural part of the translational apparatus; and proteins, the major structural and enzymatic type of molecule in cells.

10. DNA Micro Arrays

A DNA array is a collection of spots attached to a solid support such as a microscope slide where each spot contains one or more single-stranded DNA oligonucleotide fragment. Arrays make it possible to put down a large quantity of very small (100 micrometre diameter) spots on a single slide. Each spot has a DNA fragment molecule that is complementary to a single DNA sequence (similar to Southern blotting). A variation of this technique allows the gene expression of an organism at a particular stage in development to be qualified (expression profiling). In this technique the RNA in a tissue is isolated and converted to labeled cDNA. This cDNA is then hybridized to the fragments on the array and visualization of the hybridization can be done. Since multiple arrays can be made with the exact same position of fragments they are particularly useful for comparing the gene expression of two different tissues, such as a healthy and cancerous tissue. Also, one can measure what genes are expressed and how that expression changes with time or with other factors. For instance, the common baker's yeast, *Saccharomyces cerevisiae*, contains about 7000 genes; with a microarray, one can measure qualitatively how each gene is expressed, and how that expression changes, for example, with a change in temperature. There are many different ways to fabricate microarrays; the most common are silicon chips, microscope slides with spots of ~ 100 micro-meter diameter, custom arrays, and arrays with larger spots on porous membranes (macro-arrays). There can be anywhere from 100 spots to more than 10,000 on a given array.

Arrays can also be made with molecules other than DNA. For example, an antibody array can be used to determine what proteins or bacteria are present in a blood sample.

11. Fluorescence *in situ* Hybridization

Fluorescence *in situ* hybridization (FISH), a newer method for analyzing chromosomes, uses fluorescent molecules, called dyes, to "paint" genes on a chromosome. This technique is particularly useful for gene mapping and for detecting various chromosomal abnormalities. In this procedure, short sequences of DNA complementary to the sequence of interest, called probes, are hybridized to the sample DNA. Because the probes are labeled with fluorescent tags, a researcher can see the exact location of the DNA sequence of interest on a chromosome. An additional advantage of FISH is that it can be performed on non dividing cells, making it much more versatile than traditional karyotyping.

Scientists can actually create three types of FISH probes, each of which has a different application. Locus-specific probes hybridize to a particular region of a chromosome and are useful for detecting the location of a gene on a chromosome. Alphoid, or centromeric repeat probes, are generated from repetitive sequences found at the centromeres of chromosomes. Because each chromosome can be painted a different color, researchers use these probes to determine whether an individual has the correct number of chromosomes. Whole chromosome probes are actually collections of smaller probes, called libraries, that each hybridize to a different sequence along the same chromosome. Using these libraries, researchers can paint an entire chromosome with various colors, generating what is called a spectral karyotype. These types of probes are useful for examining both large- and small-scale chromosomal abnormalities.

12. Gene Mapping

Gene mapping or Genome mapping is the creation of a genetic map assigning DNA fragments to chromosomes. The term "mapping" is used in two different but related contexts. There are two different ways of mapping, *viz*.. (i) genetic mapping, and (ii) physical mapping. Genetic mapping uses classical genetic techniques (*e.g.* pedigree analysis or breeding experiments) to determine sequence features within a genome. The physical mapping uses modern techniques of molecular biology. Physical Mapping is the process of determining how DNA contained in a group of clones overlap without having to sequence all the DNA in the clones. Once the map is determined, we can use the clones as a resource to efficiently contain stretches of genome in large quantity. This type of mapping is more accurate than genetic maps.

In maps based on a genetic fingerprinting of the clones, the stretches of DNA are identified according to how they are cut by a restriction enzyme. Once cut, the DNA fragments are separated by electrophoresis. The resulting pattern of DNA migration (ie. its fingerprint) is used to identify what stretch of DNA is in the clone. By analysing the fingerprints, contigs are assembled by automated (FPC) or manual means (Pathfinders) into overlapping DNA stretches. Now a good choice of clones can be made to efficiently sequence the clones to determine the DNA sequence of the organism under study (seed picking).

Genes can be mapped prior to the complete sequencing by independent approaches like *in situ* hybridization. Once the genome has been sequenced, in-silico approaches perform the gene finding. The such suggested genes are compared with the experimental evidence for the respective gene.

13. Allele Specific Oligonucleotide

Allele specific oligonucleotide (ASO) is a technique that allows detection of single base mutations without the need for PCR or gel electrophoresis. Short (20-25 nucleotides in length), labeled probes are exposed to the non-fragmented target DNA. Hybridization occurs with high specificity due to the short length of the probes and even a single base change will hinder hybridization. The target DNA is then washed and the labeled probes that did not hybridize are removed. The target DNA is then analyzed for the presence of the probe via radioactivity or fluorescence. In this experiment, as in most molecular biology techniques, a control must be used to ensure successful experimentation.

SUMMARY

There are several molecular techniques which are widely used in molecular genetics. Important techniques of molecular genetics include (1) electrophoresis, (2) polymerase chain reaction [PCR], (3) gene cloning, (4) gene sequencing, (5) southern blot technique, (6) northern blot technique, (7) western blot technique, (8) gene silencing, (9) DNA fingerprinting, (10) DNA microarrays, (11) fluorescence *in situ* hybridization, (12) gene mapping, (13) allele specific oligonucleotides, *etc.* A brief description of all these techniques is presented.

QUESTIONS

1. **What is gene cloning? What are its uses in molecular genetics?**
2. **Define Southern, Northern and Western blot techniques and describe their important applications.**
3. **Define DNA micro arrays and describe their various applications.**
4. **What is fluorescence *in situ* hybridization? Describe its advantages.**
5. **What is DNA fingerprinting? Describe its advantages.**
6. **What are the uses of following molecular techniques?**
   (i) Gel Electrophoresis (ii) Polymerase Chain Reaction
7. **Explain briefly the following:**
   (i) Southern blot (ii) Northern blot
   (iii) Western blot (iv) DNA Fingerprinting
8. **Give the contribution of the following Scientists.**
   (i) Kary Mullis (ii) Edwin Southern
   (iii) James Alwine (iv) Alec Jeffrey
9. **Write short notes on the following:**
   (i) Gene sequencing (ii) Gene silencing
   (iii) Gene mapping (iv) Gene cloning

Chapter 2

DNA Isolation

INTRODUCTION

The DNA isolation procedure makes use of the fact that lipids and proteins are soluble in chloroform and that the DNA is soluble in water (aqueous) solutions. In plants, the DNA isolation consists of six important steps, *viz..*, grinding of tissues, centrifugation of crushed tissues, adding chloroform, rotating of solution, transfer to tube and isolation of DNA. These are discussed as follows:

1. **Grinding of Tissues:** Mechanically break pen the cell walls and membranes by grinding the tissue in a coffee grinder. The plant tissue should be frozen before grinding and ground in the coffee grinder with dry ice. Keeping the tissue frozen prevents enzymes in the cells from breaking down the DNA. [**CAUTION**: Do not touch the dry ice as it may cause "cold burns."]
2. **Centrifugation of Pulverized Tissues:** Place the crushed tissue into a 50 ml centrifuge tube containing about 10 mL of CTAB buffer solution. The buffer solution contains compounds which disable the enzymes which will break down the DNA and also a compound called CTAB which will combine with the polysaccharides and proteins and chemically alter them. Mix the solution well to suspend the cellular material. Incubate the solution at 60°C for at least 30 minutes.
3. **Adding Chloroform:** Pour an equal volume (the same amount as is already in your tube) of chloroform into the 50 ml tube. "Extract" the solution by tilting or rocking the tube gently back and forth (if you mix the solution vigorously you may break your DNA molecules!). Chloroform is an organic compound and all of the cellular compounds which are soluble in chloroform (lipids, proteins) will be dissolved into the chloroform. DNA is not soluble in chloroform and will remain dissolved in the aqueous (water) layer. [**CAUTION**: Chloroform must be used in a fume hood so that vapors will not be inhaled.]

STEPS INVOLVED IN DNA ISOLATION FROM PLANTS

↓

GRINDING OF TISSUES

↓

CENTRIFUGATION OF CRUSHED TISSUES

↓

ADDING CHLOROFORM

↓

ROTATION OF SOLUTION

↓

TRANSFER TO TUBE

↓

ISOLATION OF DNA

FIGURE 2-1: Steps Involved in DNA Isolation from Plants.

4. **Rotation of Solution:** Rotate the solution in the centrifuge for 5 minutes. The gravitational forces in the centrifuge will completely separate the two phases according to density. Chloroform is very dense and will be the bottom layer, while the water layer is less dense and will be on the top.
5. **Transfer to Tube:** Transfer the supernatant (the aqueous layer containing the DNA) to a separate tube with a pipette.
6. **Isolation of DNA:** Add 2/3 volume of ice-cold isopropanol to the aqueous layer containing the DNA [supernatant]. The DNA dissolved in the supernatant will not be soluble in the ice-cold alcohol and will start to precipitate out of solution over a period of a few minutes. After a few minutes, gently rock the tube back and forth (if you rock or shake your tube hard you will break your DNA molecules). The DNA should appear as whitish strands to clear jelly-like strands which will float in the solution.

The DNA isolation protocol for onion and wheat are outlined as follows:

Extraction of Onion DNA

Materials

1. Fresh onions

2. Graduated cylinders (10ml and 100ml)
3. Knife
4. 15 ml test tube
5. Blender
6. Test tube rack or 250 ml beaker
7. Strainer
8. Glass stirring rod
9. Coffee filters
10. Non-iodized salt
11. Adolph's natural meat tenderizer
12. Palmolive detergent
13. Beaker
14. Distilled water
15. Ice cold 95 per cent ethanol

Solutions

Detergent/Salt Solution

1. 20 ml detergent
2. 20 g non-iodized salt
3. 180 ml distilled water

5 per cent Meat Tenderizer Solution

1. 5 g meat tenderizer
2. 95 ml distilled water

Protocol

1. Cut an inch square out of the center of 3 medium onions. Chop and place in a blender.
2. Add 100 ml of detergent/salt solution.
3. Blend on high 30 sec-1 minute.
4. Strain the mixture into a beaker using a strainer with a coffee filter.
5. Add 20-30 ml meat tenderizer and stir to mix.
6. Place 6 ml filtrate in a test tube.
7. Pour 6 ml ice cold ethanol carefully down the side of the tube to form a layer.
8. Let the mixture sit undisturbed 2-3 minutes until bubbling stops.

9. The DNA will float in the alcohol. Swirl a glass stirring rod at the interface of the two layers to see the small threads of DNA.

Extraction of Wheat Germ DNA

Materials

1. 250 ml beaker
2. Baking soda
3. Hot plate
4. Adolph's natural meat tenderizer
5. Non-roasted wheat germ
6. Ice cold 95 per cent ethanol
7. Thermometer
8. 15 ml test tube
9. pH meter
10. Glass stirring rod
11. Palmolive
12. Detergent
13. Distilled water
14. Test tube rack or 250 ml beaker
15. Graduated cylinders (10ml and 100ml)

Solutions

Baking Soda Solution

1. Add baking soda to distilled water until a pH of approximately 8.0 is reached.

Protocol

1. Add 100 ml distilled water to a beaker and heat to 50-60oC.
2. Add 1.5 g wheat germ and mix until dissolved.
3. Add 5 ml detergent. Maintain 50-60oC temperature and stir for 5 minutes.
4. Add 3 g meat tenderizer.
5. Add baking soda solution to bring the pH to approximately 8.0.
6. Maintain the 50-60oC temperature and stir for 10 minutes.
7. Remove from heat.
8. Add 6 ml of the solution to a test tube and cool to room temperature.

9. Pour 6 ml ice cold ethanol carefully down the side of the tube to form a layer.
10. Let the mixture sit undisturbed 2-3 minutes until bubbling stops.
11. The DNA will float in the alcohol. Swirl a glass stirring rod at the interface of the two layers to see the small threads of DNA.

Isolation of DNA Fragments

There are two principal ways of obtaining DNA fragments, *viz*.. restriction endonuclease enzymes and Polymerase Chain Reaction. The applications of these techniques in obtaining DNA fragments are briefly discussed as follows:

1. Use of Restriction Enzymes

Before dealing with the use of restriction enzymes in obtaining DNA fragments, it is essential to define c-DNA library and genomic library. The collection of gene clones is called a DNA library. The c-DNA, or complementary DNA, is the synthetic DNA made from mRNA with the help of special enzyme called reverse transcriptase. This enzyme is isolated from retroviruses. The collection of all the c-DNA segments in suitable vectors like plasmids is known as c-DNA library. The c-DNA library is smaller than genomic library, because it contains a region of the genome whereas genomic library contain the entire genome. The enzyme reverse transcriptase synthesizes a single stranded DNA molecule using mRNA as a template. The single DNA thus synthesized can be used as a template for synthesis of double stranded DNA. The genomic library is bigger than c-DNA library, because they contain genes in their native form, including introns and regulatory sequences. For preparation of a genomic library, the total genomic DNA of an organism is extracted.

The DNA is broken into fragments of appropriate size by using a suitable restriction endonuclease enzyme or mechanical shearing or partial digestion of the DNA. For partial digestion restriction enzymes having four base recognition sequences are used. The DNA fragments produced from partially digested DNA with the help of restriction enzymes are more likely to be of appropriate size for cloning. The use of restriction enzymes has the advantage that the same set of fragments is obtained from a DNA each time. A specific enzyme is used and many of the enzymes produce cohesive ends. The partial digests of genomic DNA are subjected to agarose gel electrophoresis or sucrose gradient centrifugation for separation of fragments of appropriate size from the mixture. These fragments are then inserted into a suitable vector for cloning. This constitutes the shot gun approach for gene cloning.

Different cloning vectors carry different amounts of DNA. Hence, the choice of vector for library construction depends on the size of the genome or other DNA sample being made into the library. Plasmid and phage vectors carry small amounts of DNA. Hence, these vectors are suitable for cloning genes from organisms with small genomes. Other vectors such as Cosmids, YACs and BACs carry large amounts of DNA, hence these vectors are suitable for cloning genes from organisms with large genomes.

2. Use of Polymerase Chain Reaction [PCR]

The polymerase chain reaction (PCR) is a technique widely used in molecular biology. This technique was developed in 1985 by Kary Mullis for which she was awarded Nobel Prize in 1993. It derives its name from one of its key components, a DNA polymerase used to amplify a piece of DNA by *in vitro* enzymatic replication.

The PCR is a rapid method for cloning a particular piece of DNA in the test tube rather than in living cells like *E. coli*. In other words, PCR permits *in vitro* amplification of DNA fragments. PCR is one of the most powerful technique used in molecular biology to make multiple copies from small amount (µg microgram quantities) of desired DNA. PCR is used for gene amplification [making multiple copies of a gene] and isolation of DNA fragments of different sizes.

(i) Gene Amplification: It is possible to amplify a single or few copies of a piece of DNA through PCR and generate millions or more copies of the DNA piece. PCR can be extensively modified to perform a wide array of genetic manipulations. PCR can be used to amplify a specific gene present in different individuals of a species and even in different somatic cells or gametes, say, human sperms. These copies can be used for cloning. Sequencing of such genes will provide information on the mutational changes in the genes of different individuals, cells or gametes. Such data can be used in disease diagnosis, population genetics, estimation of recombination frequencies, *etc.*

(ii) DNA Polymorphism: PCR is widely used for isolation of DNA fragments and study DNA polymorphism in the genome using known sequences as primers. Synthetic nucleotides of any sequence can be used as random primers to amplify polymorphic DNAs having sequences specific to the primers used. PCR generates random amplified polymorphic DNA' (RAPD), which is detected as bands after electrophoresis. RAPD bands of different strains or species can be compared. They can be used to construct RAPD maps, similar to RFLP maps. In plants, several other DNA markers can be generated through PCR such as AFLP, SSRs, SNP, *etc.*

Gene Detection

The presence of gene of interest in a sample can be detected by determining the presence of three components, *viz..*, DNA, mRNA and protein. There are three main techniques detect the presence of gene of interest in a sample. These are southern blotting, northern blotting and western blotting. For details of these blotting techniques please refer Chapter 5.

SUMMARY

DNA is isolated from the rest of the cell contents by taking advantage of the different chemical and physical properties of the molecules found within the cell. Each organism is different and the cells of each kind of organism often contain special compounds, which must be extracted away from the DNA.

The DNA isolation procedure makes use of the fact that lipids and proteins are soluble in chloroform and that the DNA is soluble in water (aqueous) solutions. In plants, the DNA isolation consists of six important steps, *viz..*, grinding of tissues, centrifugation of crushed tissues, adding chloroform, rotating of solution, transfer to tube and isolation of DNA. These have been discussed. The DNA isolation protocol for onion and wheat are outlined.

There are two principal ways of obtaining DNA fragments, *viz..* restriction endonuclease enzymes and Polymerase Chain Reaction. The applications of these techniques in obtaining DNA fragments have been discussed.

The presence of gene of interest in a sample can be detected by determining the presence of three components, *viz..*, DNA, mRNA and protein. There are three main techniques detect the presence of gene of interest in a sample.

QUESTIONS

1. **Describe briefly various steps involved in DNA isolation protocol.**
2. **Describe DNA isolation protocol in onion.**
3. **Explain DNA isolation protocol in wheat seeds.**
4. **Discuss methods of extraction of DNA fragments.**
5. **Describe the role of Polymerase Chain Reaction in isolation of DNA fragments.**
6. **How gel electrophoresis is useful in isolation of DNA fragments?**
7. **Write short notes on the following:**

   (i) Gel electrophoresis (ii) Polymerase Chain Reaction

Chapter 3

Gene Cloning

INTRODUCTION

The process by which multiple copies of a specific gene (DNA) can be made after isolation of such gene or DNA is known as gene cloning. In other words, gene cloning is the process of making several copies of a gene. Gene cloning is known by a variety of names such as DNA cloning, molecular cloning and recombinant DNA technology. In gene cloning, a segment of DNA representing a gene is inserted in to a cloning vector where it replicates further. Main points related to gene cloning are listed as follows:

1. Gene cloning is also known as DNA cloning or genetic cloning or molecular cloning or gene amplification.
2. Gene cloning is done using bacterial plasmid [Ti plasmid].
3. It involves an individual gene of interest.
4. It is also referred to as recombinant DNA (r-DNA) technology.
5. Gene cloning is a pre-requisite for genetic transformation.

In gene cloning, an individual gene is isolated, amplified and inserted in a bacterium for further replication. Gene amplification refers to an increase in the copy number of a gene or DNA sequence. The recombinant DNA [r-DNA] technique is often used as synonym for DNA cloning or gene cloning. The construction of composite (or) artificial DNA molecule is termed as gene manipulation or r- DNA technique or sometimes as genetic Engineering.

Gene cloning is the process in which a gene of interest is located and copied (cloned) out of DNA extracted from an organism. When DNA is extracted from an organism, all of its genes are extracted at one time. From this DNA, which contains thousands of different genes, the genetic engineer must find the one specific gene that encodes the specific protein of interest.

Steps Involved

Gene cloning is the method of producing identical genes through different procedure. The gene cloning consists of five important steps, *viz..*, isolation of the desired gene, introduction of desired gene into the vector, insertion of plasmid into the host cell, cloning of new gene and isolation of the cloned gene. These steps are briefly discussed as follows:

1. Isolation of the Desired Gene

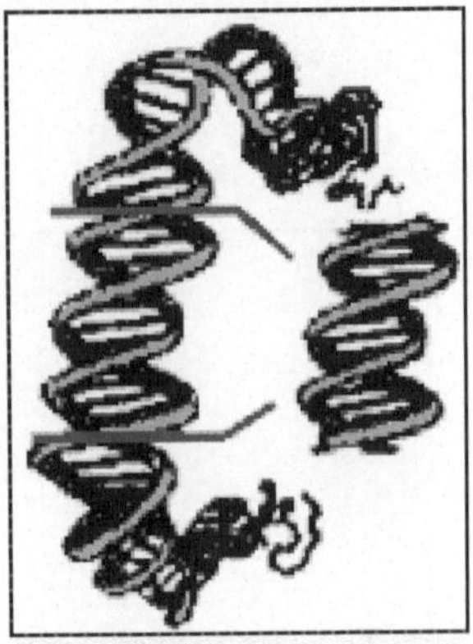

The process of gene cloning.

First step in gene cloning is isolation of the desired part of DNA in which the gene of interest is present. Similarly, the bacterial plasmids are also isolated in which the desired gene will be inserted. When the desired region with gene of interest is identified, it is isolated by cutting it with the help of restriction enzymes. As all enzymes are proteins, restriction enzymes are also proteins which play a role of catalyst in chemical reaction. They cut the phosphodiester bonds, a covalent bond of deoxyribose in DNA. There are specific sites on the DNA molecule called restriction sites. These are the locations where from gene of interest is isolated from the DNA molecule. The desired gene is isolated by breaking the strand of DNA. Now this gene is ready for introduction to the vector.

First of all, the gene of interest is isolated from chromosomal DNA using restriction enzyme. The restriction enzyme is used to isolate single gene from the entire genome.

2. Introduction of the Desired Gene into the Vector

After isolation, the gene of interest is inserted in to a host cell where it can replicate and make multiple copies. For this purpose, a plasmid is used which is extra chromosomal element inside the bacterial cell. It has the ability to replicate independently of the bacterial chromosomal DNA. Plasmids are the best source of gene cloning as they are able to replicate separately and independently of bacteria's own genetic material. They are double stranded circular molecules. When the gene of interest is inserted into the plasmid, the ring of the plasmid opens up giving place to the gene to attach with it. Now the plasmid is ready for the introduction into the host organism as it contains foreign gene.

Both the gene of interest and plasmid are treated with same restriction enzyme to produce identical sticky ends. The isolated gene or DNA fragment is inserted into a suitable vector (plasmid) to produce recombinant DNA (rDNA). The DNAs from both sources are mixed together and treated with enzyme ligase to unite them together. When the fragmented chromosomal DNA (gene of interest) is joined with cloning vector in the laboratory, it is called recombinant DNA molecule. Thus, a r-DNA molecule is produced by joining together two or more DNA segments usually originating from different organisms. The r-DNA molecule is a vector (a plasmid) into which the gene of interest has been inserted for its cloning in an appropriate host. In this process, restriction enzymes are used for cutting of DNA into suitable fragments and ligase enzyme is used for joining together the appropriate fragments.

3. Insertion of Plasmid into the Host Cell

The *Escherichia coli,* species of bacteria useful for human body, is the most commonly used plasmid. The plasmid with foreign gene is now inserted into host organism's body. Plasmid enters the host cell by two mechanisms. Firstly it is inserted by placing the cell in calcium chloride which will enable the plasmid to enter the host cell. Secondly, it is inserted by electroporation in which slight electric current is applied resulting in formation of pores in the host cell's membrane. The plasmid can enter through these pores. Now the host cell is genetically modified, because it contains foreign gene.

Then r-DNA (vector) is introduced into a suitable host cells (bacterial cells) for reproduction along with the host cell DNA. Generally, *E. coli* is used for initial cloning. As the bacterial cells reproduce (by mitosis), the recombinant plasmid with gene of interest is copied. Thus millions of copies of the gene are produced by gene cloning.

STEPS INVOLVED IN GENE CLONING

↓

ISOLATION OF DESIRED GENE

↓

INTRODUCTION INTO THE VECTOR

↓

INSERTION OF PLASMID IN HOST CELL

↓

CLONING OF NEW GENE

↓

ISOLATION OF CLONED GENE

FIGURE 3-1: Steps Involved in Gene Cloning.

4. Cloning of New Gene

Since plasmid is independent of chromosomal DNA, after insertion into the bacterial cell, it starts replicating at a very high speed making identical copies of desired gene. Scientists use *E. coli* as the host cell for the plasmid, because it has the ability to replicate faster than any other organism. It is the best micro-organism used in recombinant DNA technology.

5. Isolation of the Cloned Gene

After cloning, these bacterial cells are put in to the culture, where they replicate. Then the cloned gene is isolated again by using restriction enzymes through the process known as lyses. In this way, the cloned gene can be isolated without any damage to the gene.

The r-DNA molecule in which the coding region from one organism is joined to regulatory sequence of another organism is called chimeric gene. The capability to produce r-DNA molecules has given human beings the power and opportunity to create novel gene functions to suit specific needs. Recombinant DNA molecule is produced to obtain multiple copies of specific DNA fragments, recover large quantities of the protein production by the concerned gene and integrate gene into the chromosome of target organism for genetic transformation.

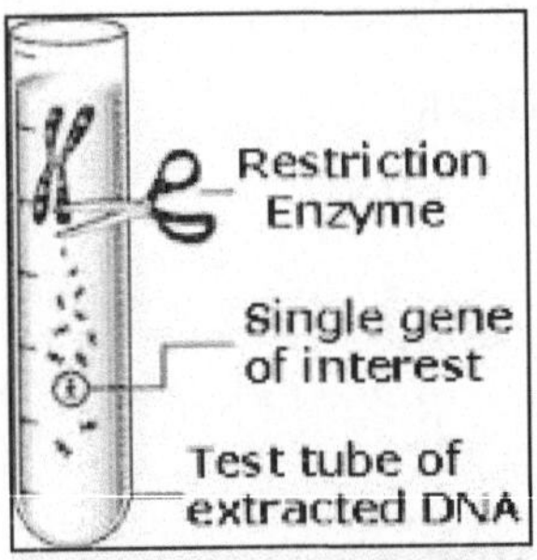

Restriction enzymes cut the extracted DNA sequences into gene-size pieces.

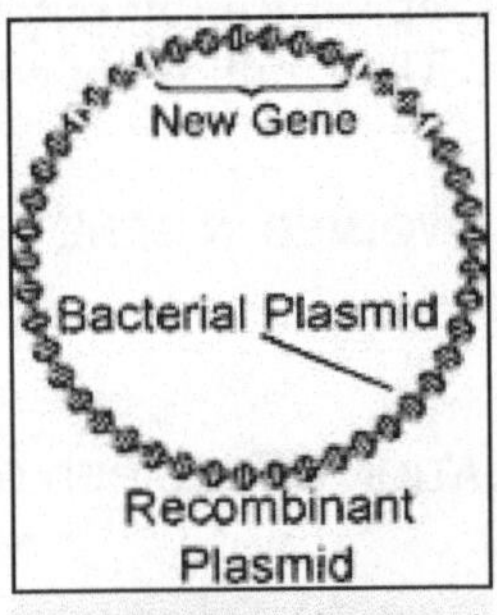

Bacterial plasmids are cut with the same restriction enzymes. They are then combined with the extracted DNA and the two often anneal together.

After undergoing electroporation or heat shock, the bacteria are plated out onto petri dishes.

Gene Libraries

The collection of gene clones is called a DNA library or gene library. The gene libraries are made to catalogue the DNA of an organism. The gene of interest is selected from the gene library. The library construction starts after DNA extraction of an organism. If a library is to have a colony of bacteria for every gene, it will consist of thousands of colonies or clones.

Construction of Library

Constructing a gene library requires extracted DNA, restriction enzymes and a plasmid. The construction of gene library consists of several steps which are explained as follows:

1. **Cutting of DNA into Gene Size:** DNA extracted from an organism, with the gene of interest, is cut into gene-size pieces with restriction enzymes. These enzymes read the nucleotide sequence of the DNA and recognize

specific sequences. The enzymes then cut the DNA sequence by breaking the bonds between nucleotides in a DNA strand.

2. **Cutting of Bacterial Plasmid:** Bacterial plasmids are cut with the same restriction enzyme. Plasmids are small circles of DNA in bacterial cells that are naturally present in addition to the bacteria's other DNA.
3. **Combining Gene and Plasmid:** The gene-sized DNA and cut plasmids are combined into one test tube. Some of the enzyme cut DNA pieces will combine with the cut plasmids and form recombinant DNA (or DNA in a 'new combination').
4. **Transfer of Plasmid into Bacteria:** The recombinant plasmids are then transferred into bacteria using either electroporation or heat shock. Electroporation uses mild pulses of electricity to disrupt the cell walls of the bacterium and create small holes. The plasmids are small enough to pass through the holes into the cell. Heat shock works in a similar fashion. However, rather than using electricity to create holes in the bacterium, it is done by alternating the temperature between hot and cold.
5. **Growing of Bacteria in Culture:** The bacteria is grown on a culture dish and allowed to grow into colonies. All the colonies on all the plates (cultures) are called a gene library.
6. **Screening of Gene Library:** The gene library is then screened in order to discover which bacterial colony is making copies of the one gene they are interested in. Library screening identifies colonies, which have that particular gene. Screening can be based on detecting the DNA sequence of the cloned gene, detecting a protein that the gene encodes, or the use of linked DNA markers. Therefore before library screening can be done, the scientist must know either the DNA sequence of the gene, or a very similar gene, the protein that the gene produces, or a DNA marker that has been mapped very close to the gene. When the bacteria multiply and replicate the recombinant DNA, the number of gene copies also increases, making gene or protein detection easier.

When the bacteria colony containing the desired gene is located, the bacteria can be propagated to make millions of copies of the recombinant plasmid that contains the gene. The plasmids can be extracted for the next steps of genetic engineering, gene modification, and transformation. Gene cloning is also important because copies of a gene are needed for these procedures.

Advantages of Gene Cloning

There are several advantages of gene cloning. Gene cloning is useful in determining gene nucleotide sequence, gene evolution, gene function, development of transgenic plants, production of human insulin. These are explained as follows:

1. **Determining Nucleotide Sequence:** Isolation of individual gene helps in determining its nucleotide sequence. This will provide information about intron number and position and promoter elements.

2. **Gene Evolution:** A comparison of DNA sequences between genes will provide information about gene evolution.
3. **Gene Function:** Gene cloning willprovide information about the function of individual gene. This can be easily determined from the structure of protein produced by a gene.
4. **Development of Transgenics:** The gene cloning is useful in developing transgenic crops with better quality and resistance to insects, diseases and herbicides.
5. **Production of Insulin:** Gene cloning is useful in developing transgenic bacteria which can be used for producing human insulin using appropriate human gene.

Limitations of Gene Cloning

There are some limitations of gene cloning. The main limitations include time and labor, well equipped laboratory, well trained persons and adequate funds. These are explained as follows:

1. **Time Consuming:** Gene cloning requires lot of time and labor. Thus it is a cumbersome job.
2. **Sophisticated Laboratory:** Gene cloning requires well equipped sophisticated laboratory, costly glassware and chemicals.
3. **Technical Skill:** Gene cloning also requires a team of well trained persons.
4. **Adequate Funds:** Adequate funds are required for smooth functioning of gene cloning laboratory which often becomes a limiting factor.

Application in Agriculture

Bacteria are also important for the nitrogen fixation Bacteria along with desired gene are used to increase crop productivity and health. They make farmers free of using expensive fertilizers which can damage the crop.

SUMMARY

Gene cloning is a process by which multiple copies of a specific gene (DNA) can be made after isolation of such gene or DNA. In other words, gene cloning is the process of making several copies of a gene. Gene cloning is known by a variety of names such as DNA cloning, molecular cloning and recombinant DNA technology.

Gene cloning is the method of producing identical genes through different procedure. The gene cloning consists of five important steps, *viz..*, isolation of the desired gene, introduction of desired gene into the vector, insertion of plasmid into the host cell, cloning of new gene and isolation of the cloned gene. These steps have been briefly discussed.

There are several advantages of gene cloning. Gene cloning is useful in determining gene nucleotide sequence, gene evolution, gene function, development of transgenic plants, production of human insulin. These have been explained.

There are some limitations of gene cloning. The main limitations include time and labor, well equipped laboratory, well trained persons and adequate funds. These are explained as follows:

QUESTIONS

1. **Define gene cloning and describe briefly various steps involved in gene cloning.**
2. **What are advantages of gene cloning?**
3. **What are limitations of gene cloning?**
4. **What are the steps involved in the construction of gene library?**
5. **Discuss applications of gene cloning in agriculture.**
6. **Write short notes on the following:**
   (i) Gene cloning (ii) Gene library
   (ii) Genomic library (iv) Cloned gene

Chapter 4

Polymerase Chain Reaction (PCR)

INTRODUCTION

The polymerase chain reaction technique was developed by Kary Mullis in 1985 for which she was awarded Nobel Prize in 1993. The polymerase chain reaction (PCR) is a technique which is widely used in molecular biology. It derives its name from one of its key components, a DNA polymerase used to amplify a piece of DNA by *in vitro* enzymatic replication. The main points related to PCR are listed as follows:

1. **Rapid Method:** The PCR is a rapid method of cloning a particular piece of DNA in the test tube rather than in living cells like *E. coli*. In other words, PCR permits *in vitro* amplification of DNA fragments.
2. **Powerful Technique:** PCR is one of the most powerful technique used in molecular biology to make multiple copies from small amount (µg microgram quantities) of desired DNA. With PCR, it is possible to amplify a single or few copies of a piece of DNA in to millions or more copies.
3. **Use of Heat Resistant Enzyme:** Almost all PCR applications employ a heat-stable DNA polymerase, such as Taq polymerase, an enzyme originally isolated from the bacterium *Thermus aquaticus*. This DNA polymerase enzymatically assembles a new DNA strand from DNA building blocks, the nucleotides, using single-stranded DNA as template and DNA oligonucleotides (also called DNA primers) required for initiation of DNA synthesis.
4. **Use of Thermal Cycling:** The vast majority of PCR methods use thermal cycling, *i.e.*, alternately heating and cooling the PCR sample to a defined series of temperature steps. These thermal cycling steps are necessary to physically separate the strands (at high temperatures) in a DNA double helix (DNA melting) used as template during DNA synthesis (at lower temperatures) by the DNA polymerase to selectively amplify the target DNA.

5. **Use of Primers:** The primer is a short sequence of RNA) or an oligonucleotide that pairs with one strand of template DNA and provides a 3′ OH essential for DNA polymerase to extend the primer through DNA synthesis.
6. **Principle Involved:** PCR is based on the principle of semi-conservative DNA replication carried out by DNA polymerase in prokaryotic and eukaryotic cells.
7. **Ingredients Required:** The 'ingredients' needed in PCR include a segment of the DNA strand to be analyzed, free nucleotides containing the nitrogen bases [adenine, thymine, guanine and cytosine], synthesized 'primer' sections of DNA, a heat stable DNA polymerase such as Taq polymerase and a primase enzyme that anneals the primer sequences to the unzipped DNA strands. The nitrogen bases are required in the form of deoxynucleotide triphosphates.

Steps Involved

DNA amplification by the PCR is carried out in three main steps, *viz..* denaturation, annealing and primer extension or synthesis. These are repeated for a number of cycles to increase the number of copies of a specific target region of the DNA. Three steps are explained as follows:

1. **Denaturation:** The separation of two strands of DNA in to single strands is called denaturation. The genomic DNA is normally double stranded. For gene amplification single stranded DNA is required. The DNA double helix is separated into two single strands by heating reaction between 90-98°C. The duration of this step in the first cycle of PCR is usually, 2 min at 94°C, but in subsequent cycles it is of only 1min duration.

   The PCR process commences by heating the reaction mixture to 95°C. This denatures the DNA sample to form two separate strands. The temperature of the mixture is then reduced to 60°C, to provide ideal conditions for the primer sequences to form hydrogen bonds with their complementary sections on the free DNA strands.
2. **Annealing:** Annealing refers to the process of hybridization of a mixture of DNAs in single stranded form and detection of homologous regions in the DNA of two species involved. The mixture is now cooled to a temperature of generally 40-60°C that permits annealing of primer to the complimentary sequences in the DNA. The duration of annealing step is usually 1 min during the first as well as the subsequent cycles of PCR. The annealed primers are then extended (*i.e.* synthesis of DNA) with Taq DNA polymerase.
3. **Primer Extension (or) synthesis:** It involves heating the mixture to 72°C at which a special polymerase synthesizes the new DNA strand by using the original strand as the template starting by the primer utilizing their 31OH free ends and continuing in the 31 direction. The first cycle of

amplification is over and these three steps are repeated 25-40 times to produce millions of exact copies of the target region of DNA. Thus at the end of each cycle the number of copies of desired segment becomes twice the number present at the end of previous cycle.

In case of automated PCR machines, called thermal cycles, the researcher specify only the number and duration of cycle *etc.* after placing the complete reaction mixture for incubation, and the machine performs the entire operations preciously. After PCR cycles the amplified DNA segment is purified by gel electrophoresis and can be used for cloning, DNA sequencing, *etc.* As PCR progresses, the DNA thus generated is itself used as template for replication. This sets in motion a chain reaction in which the DNA template is exponentially amplified.

TABLE 4-1: Comparison of PCR and Gene Cloning Techniques

Sl.No.	*Particulars*	*PCR*	*Gene Cloning*
1	Cost involved	Less	More
2	Technical skill	Not required	Required
3	Applications	Wide	Less
4	Accuracy	High	Less
5	Automation	Possible	Not possible
6	Speed of technique	High	Low
7	Labor required	Less	More
8	Manipulation	*In vitro*	*In vitro* and *In vivo*
9	Primers	Required	Not required

Applications of PCR

Polymerase Chain Reaction [PCR] has several useful applications. Some important applications of PCR include gene amplification, study of DNA polymorphism, detection of transgenes, location of genes, detection of sex and detection of paternity. These are briefly discussed as follows:

1. Gene Amplification

It is possible to amplify a single or few copies of a piece of DNA through PCR and generate millions or more copies of the DNA piece. PCR can be extensively modified to perform a wide array of genetic manipulations. PCR can be used to amplify a specific gene present in different individuals of a species and even in different somatic cells or gametes, say, human sperms. These copies can be used for cloning. Sequencing of such genes will provide information on the mutational changes in the genes of different individuals, cells or gametes. Such data can be used in disease diagnosis, population genetics, estimation of recombination frequencies, *etc.*

2. DNA Polymorphism

PCR has been used to study DNA polymorphism in the genome using known sequences as primers. Synthetic nucleotides of any sequence can be used as random primers to amplify polymorphic DNAs having sequences specific to the primers used. PCR generates random amplified polymorphic DNA' (RAPD), which is detected as bands after electrophoresis. RAPD bands of different strains or species can be compared. They can be used to construct RAPD maps, similar to RFLP maps. Several other DNA markers can be generated through PCR.

3. Detection of Transgenes

Polymerase Chain Reaction [PCR] can be used to detect the presence of transgene introduced into an organism by using the end sequences of the transgene for amplification of DNA from the putative transgenic organism. Amplification will occur only when the transgene is present in the organism. The amplified DNA is detected as a band on the electrophoretic gel.

4. Location of Genes

The physical location of specific genes in chromosomes can be determined through the PCR amplification of micro-dissected segments of chromosomes, *e.g.*, of salivary gland chromosomes of *Drosophila.*

5. Detection of Sex

Polymerase Chain Reaction [PCR can be used to determine the sex of embryos. Thus the sex of *in vitro* fertilized cattle embryos could be determined using Y chromosome specific primers before their implantation in the uterus.

6. Detection of Paternity

The paternity of disputed children can be easily decided through identification of genetic finger prints. The can also be used in the detection of criminals.

Advantages and Disadvantages of PCR

There are some advantages and some disadvantages or limitations of PCR which are briefly presented as follows:

Advantages

1. It can create multiple copies from a small sample.
2. It is good for criminal investigations even with some hair.
3. PCR is a rapid technique. It can make millions of copies in few hours.

Disadvantages

PCR cannot substitute for gene cloning in cells when large amounts of a gene are desired.

1. Also occasional errors during PCR replication impose limits on the number of good copies that can be made.
2. PCRs of longer products are less efficient due to enzyme activity loss, and inaccuracies introduced by longer PCRs.

3. Mistakes and errors do happen in PCR by DNA polymerases in the test tube and in the human body.
4. The polymerases used in PCR often lack 3′ to 5′ exonuclease activity such as Taq polymerase. This enzyme lacks the ability to correct mis-incorporated nucleotides, an activity which is present in higher eukaryotes.

SUMMARY

The polymerase chain reaction (PCR) is a technique widely used in molecular biology. This technique was developed in 1985 by Kary Mullis for which she was awarded Nobel Prize in 1993. It derives its name from one of its key components, a DNA polymerase used to amplify a piece of DNA by *in vitro* enzymatic replication.

DNA amplification by the PCR is carried out in three main steps, *viz*.. denaturation, annealing and primer extension or synthesis. These are repeated for a number of cycles to increase the number of copies of a specific target region of the DNA. Three steps have been explained.

Polymerase Chain Reaction [PCR] has several useful applications. Some important applications of PCR include gene amplification, study of DNA polymorphism, detection of transgenes, location of genes, detection of sex and detection of paternity. These have been briefly discussed. There are some advantages and some disadvantages or limitations of PCR which have been listed.

QUESTIONS

1. **What is PCR? Describe briefly main features of PCR.**
2. **Explain briefly the various steps involved in gene amplification by PCR.**
3. **What are various applications of PCR?**
4. **Describe briefly advantages and disadvantages of PCR.**
5. **Give a brief comparison of PCR and gene cloning techniques.**
6. **Explain briefly the role of PCR in the detection of transgenic plants and sex**
7. **Describe the main disadvantages of PCR technique**
8. **What are the advantages of PCR as compared to restriction enzymes?**
9. **Describe PCR in relation to:**

   (i) Gene amplification
   (ii) DNA polymorphism
   (iii) Denaturation
   (iv) Annealing

Chapter 5

Gene Detection Techniques

INTRODUCTION

The presence of gene of interest in a sample can be detected by determining the presence of three components, *viz..*, DNA, mRNA and protein. There are three main techniques which are used to detect the presence of gene of interest in a sample. These are southern blotting, northern blotting and western blotting. All these techniques make use of gel electrophoresis. Hence, brief description of gel electrophoresis is essential before dealing with gene detection techniques.

Gel Electrophoresis

Gel electrophoresis is one of the principal tools of molecular biology. The basic principle is that DNA, RNA, and proteins can all be separated by means of an electric field. In agarose gel electrophoresis, DNA and RNA can be separated on the basis of size by running the DNA through an agarose gel. Proteins can be separated on the basis of size by using an SDS-PAGE gel, or on the basis of size and their electric charge by using what is known as a 2D gel electrophoresis. There are many forms of electrophoresis in use but all operate by putting a solution of the molecules to be separated into an electric field. In all electrophoresis experiments electrodes are positioned at opposite ends of the separation matrix which may be in the form of a tube or of a thick or thin slab. In the most common forms of electrophoresis two type of matrices, *viz..* agarose and polyacrylamide are used. There are many different commercial and homemade apparatuses in use. For unimportant technical reasons, acrylamide gels are usually held vertically and the samples are loaded in wells formed in the top edge whereas agarose gels are usually held horizontally and the samples are inserted into wells in their top face.

Southern Blotting

A Southern blot is a method routinely used in molecular biology to detect the presence of a specific DNA sequence (a gene) in a DNA sample. This technique was developed by Edwin Southern in 1975. Main points related to southern blotting technique are listed below.

1. **Other Term Used:** Southern blotting is also known as southern hybridization, southern transfer or DNA gel blot analysis.
2. **Uses:** it is used for detecting specific DNA fragments in a complex mixture. of DNA sample.
3. **Labels Used:** Most original protocols used radioactive labels. However, non-radioactive alternatives are now available. The labeled DNA is used as a probe for easy identification. The probe is labeled either with radioactive chemical or fluorescent dye.
4. **Restriction Enzyme:** Restriction enzyme is used in this technique to cut the DNA into fragments. DNA samples before or after restriction enzyme digestion are separated by gel electrophoresis and then transferred to a membrane by blotting via capillary action.
5. **Gel Used:** In this technique, agarose gel is used for DNA detection.
6. **Membrane Used:** In this technique nitrocellulose or nylon membrane is used. The membrane is then exposed to a labeled DNA probe that has a complementary base sequence to the sequence on the DNA of interest.
7. **Blotting Method:** The DNA is separated by gel electrophoresis and then transferred to a membrane by blotting via capillary action.
8. **Denaturant Used:** In this technique sodium hydroxide is used as denaturant.
9. **Applications:** Southern blotting is less commonly used in laboratory science due to the capacity of other techniques, such as PCR, to detect specific DNA sequences from DNA samples. These blots are still used for some purposes such as detection of transgene and their copy number in transgenic organisms.

Protocol of Southern Blotting

This technique was developed by E.M Southern in 1975 for the transfer of DNA fragment from an electrophoresis gel to the nitrocellulose filter or nylon membrane by capillary action. It involves DNA-DNA hybridization and consists of the following sequential steps.

1. **Isolation and Purification of DNA**: The DNA is isolated from the cells or tissues of interest using standard DNA isolation protocol. The DNA isolation protocols differ from species to species to some extent. After isolation, the DNA is purified.
2. **Cutting of DNA into Fragments:** The DNA is digested with an appropriate restriction enzyme to cut it into fragments of many size. It can be started with intact chromosomes also.

3. **Separation of Fragments by Size:** The digested or fragmented DNA is separated into fragment of different sizes through gel electrophoresis using agarose gel.
4. **Denaturation of ds DNA:** Then the double stranded DNA is denatured into single strands by alkali treatment *i.e.* incubation with sodium hydroxide (NaOH). This will allow complementary base pairing with the probe.
5. **Transfer onto membrane:** Then DNA fragments are transferred from gel onto the membrane by capillary action. This is called blotting. Two types of membranes, *viz..*, nitrocellulose membrane and nylon membrane are used for providing solid support to the denatured DNA.
6. **Incubation with a labeled Probe:** A brobe refers to a small segment of DNA with known base sequences, origin and function. The probe is complementary to the gene of interest. The probe is labeled with radioactive chemical (32P) or fluorescent dye for easy identification. The probe will bind strongly to the sequences in the genome which show large amount of sequence similarity to the probe. It will lead to formation of double stranded DNA due to hybridization of probe with genomic DNA. This is called southern hybridization. Thus the probe hybridizes with complimentary DNA on the membrane resulting in DNA-DNA hybridization.
7. **Washing and Detection:** The washing is done to remove unbound or poorly bound molecules. After washing only hybrid (tagged) molecules are left. The hybrid sequence having one labeled DNA strand can be easily detected. The autoradiography or X-ray film reveals the positions of DNA segments in the gel which are complimentary to the radioactive probes.

Northern Blotting

The northern blot is a technique used in molecular biology research to determine the presence of a specific mRNA from a set of different samples of RNA. This technique was developed in 1977 by James Alwine, David Kemp, and George Stark at Stanford University and further elaborated by Thomas in 1980. It takes its name from the similarity of the Southern blot technique. Main points related to this technique are listed as follows:

1. **Other Term Used:** Northern blotting is also known as northern transfer or RNA gel blot analysis.
2. **Uses:** This is used for detection of RNA in a sample. In the northern blot, RNA, rather than DNA, is analyzed by electrophoresis and detected with a hybridization probe. In this process RNA is separated based on size and is then transferred to a membrane that is then probed with a labeled complement of a sequence of interest.
3. **Labels Used:** The labeled DNA is used as a probe for easy identification. The probe is labeled either with radioactive chemical or fluorescent dye.

4. **Restriction Enzyme:** Restriction enzyme is not used in this technique.
5. **Gel Used:** The gels may be run on either agarose or denaturing polyacrylamide gels depending on the size of the RNA to be detected. The intensity of bands is related to the amount of the target RNA in the samples analyzed.
6. **Membrane Used:** In this technique nitrocellulose or nylon membrane is used.
7. **Blotting Method:** The RNA is separated by gel electrophoresis and then transferred to a membrane by blotting via capillary action.
8. **Denaturant Used:** In this technique, formaldehyde is added in agarose gel which acts as a denaturant. For smaller fragments denaturing polyacrylamide urea gels are employed.
9. **Applications:** It can be used for DNA finger printing, preparation of RFLP maps, detection and identification of transferred genes in the transgenic individuals *etc.*

TABLE 5-1: Comparison of Southern, Northern and Western Blotting Techniques

Sl.No.	*Particulars*	*Southern Blotting*	*Northern Blotting*	*Western Blotting*
1	Invented by	Edwin Southern	Alwine *et al.*	Towbin *et al.*
2	Year of invention	1975	1977	1979
3	Other term used	Southern hybridization	Northern hybridization	Immunoblot
4	Used for detection of	DNA	RNA	Protein
5	Probe used	Labeled DNA	Labeled DNA	Antibody
6	Restriction enzyme	Used	Not used	Not used
7	Membrane used	Nitrocellulose or nylon	Nitrocellulose or nylon	Nitrocellulose or nylon
8	Antibody	Not used	Not used	Used
9	Denaturant used	NaOH	Formaldehyde	-
10	Positive control indicates presence of	DNA of interest	RNA of interest	Protein of interest
11	Gel used	Agarose gel	Formaldehyde agarose gel	Polyacrylamide gel
12	Blotting method	Capillary transfer	Capillary transfer	Electric transfer

Western Blotting

This technique was developed by Towbin and coworkers in 1979 which originated from the laboratory of George Stark at Stanford. The western blot is a method of detecting the presence of a specific proteins in a given sample of tissues or extract using antibody specific to that protein. The main points related to this technique are listed as follows:

1. **Other Term Used:** It is also known as Western blot or immune blot technique.
2. **Uses:** Western blot is used for detecting a specific proteins in a given sample of tissues or extract.
3. **Electrophoresis:** It uses gel electrophoresis to separate native or denatured proteins by the length of the polypeptide (denaturing conditions) or by the 3-D structure of the protein. In this technique, proteins are first separated by size, in a thin gel sandwiched between two glass plates in a technique known as SDS-PAGE (sodium dodecyl sulfate polyacrylamide gel electrophoresis). Protein samples are separated by gel electrophoresis and then transferred to a membrane by blotting via capillary action.
4. **Label Used:** DNA probe is not used. An antibody is used as a probe. Antibodies that specifically bind to the protein of interest can then be visualized by a variety of techniques, including colored products, chemiluminescence, or autoradiography. Often, the antibodies are labeled with an enzyme.
5. **Restriction Enzyme:** In this technique, restriction enzyme is not used.
6. **Gel Used:** In this technique, polyacrylamide gel is used for protein detection.
7. **Membrane Used:** In this technique nitrocellulose or nylon membrane is used. The proteins in the gel are then transferred to a PVDF, nitrocellulose, nylon or other support membrane. This membrane can then be probed with solutions of antibodies. When a chemiluminescent substrate is exposed to the enzyme it allows detection.
8. **Blotting Method:** separated by gel electrophoresis and then transferred to a membrane by blotting via electric action.
9. **Applications:** This method is used in the fields of molecular biology, biochemistry, immunogenetics and medical sciences. Other related techniques include using antibodies to detect proteins in tissues and cells by immunostaining and enzyme-linked immunosorbent assay (ELISA). In medical science, western blot technique is used as (i) confirmatory test for HIV, (ii) definite test for Bovine spongiform encephalopathy (BSE, commonly referred to as 'mad cow disease'), and (iii) definitive test for some forms of Lyme disease testing. This technique not only allows detection of specific protein but also quantitative analysis.

Probes for DNA Detection

A small nucleotide sequence [15-30bp] of DNA or RNA which is used to detect the presence of complimentary sequences of nucleic acid in a sample is referred to as probe. For easy identification in autoradiography of the DNA samples, the probes are usually labeled with p32. Single stranded probes are desirable to avoid pairing between the two strands of the probe itself. A probe hybridizes with the segment having its complementary nucleotide sequences in the sample to be assayed.

Both the DNA and RNA sequences are used as probes which can be obtained from either genomic DNA or c-DNA libraries. Another source of DNA for a probe might be homologous gene from a related organism. DNA probe can be synthesized if the protein product of gene of interest is known.

Applications

DNA probes are useful in identification of r-DNA clones, plant varieties, transgenes, criminals and paternity; isolation of DNA segment; development of RFLP maps and genomic maps of eukaryotes and disease diagnosis.

SUMMARY

The presence of gene of interest in a sample can be detected by determining the presence of three components, *viz..*, DNA, mRNA and protein. There are three main techniques detect the presence of gene of interest in a sample. These are southern blotting, northern blotting and western blotting. These techniques have been amply discussed. A comparison of these techniques has been given in tabular form.

A small nucleotide sequence [15-30bp] of DNA or RNA which is used to detect the presence of complimentary sequences of nucleic acid samples is referred to as probe. Both the DNA and RNA sequences are used as probes which can be obtained from either genomic DNA or c-DNA libraries.

DNA probes are useful in identification of r-DNA clones, plant varieties, transgenes, criminals and paternity; isolation of DNA segment; development of RFLP maps and genomic maps of eukaryotes and disease diagnosis.

QUESTIONS

1. **What are gene detection techniques? Describe any one of them in detail.**
2. **Define southern blotting. Explain various aspects of this technique.**
3. **What is northern blotting? Explain various aspects of this technique.**
4. **Define Western blotting. Describe various aspects of this technique.**
5. **Give a brief comparison of Southern blot, Northern blot and Western blot techniques.**
6. **Define DNA probe and describe their applications.**
7. **Define the following terms**

   (i) Southern blotting   (ii) Northern blotting

   (iv) Western blotting   (iv) DNA probes

Chapter 6

Genotyping and Phenotyping

INTRODUCTION

Terms genotype and phenotype were first used by Johannsen, a Danist Biologist, in in 1909 working with common bean (*Phaseolus vulgaris*). The genetic constitution or make up of an individual is known as genotype. In other words, genotype is the structure of DNA which determines the expression of a trait. Thus molecular composition of an individual is known as genotype. Main features of genotype are listed below:

(i) It refers to molecular composition of an individual.
(ii) Individuals with homozygous genotypes will always breed true on selfing.
(iii) For a single gene difference, the genetic ratio in F2 generation will be 1:2:1.
(iv) Genotypic classes are always higher than phenotypic classes.
(v) Genotype is not influenced by environmental factors.
(vi) Genotypes can be observed at molecular level only.
(vii) A genotype is composed of hundreds of genes or alleles.
(viii) A genotype expresses through phenotype. For example a red allele [RR] genotype will produce red coloured phenotype.
(ix) A genotype can be altered by mutations and hybridization.
(x) A genotype may exhibit additive or non-additive type of gene interaction.

Phenotype refers to external appearance of an individual. In other words, it refers to visual characteristics of an individual such as colour, height, sex, *etc.* For example, purple and white flower colours are two different phenotypes. Main features of phenotypes are listed as follows:

(i) It refers to observable characteristics of an individual such as flower colour, leaf shape, surface and stature of plants.

(ii) Phenotype exhibits outward manifestation of a specific genotype.

(iii) Phenotype consists of hundreds of morphological physiological and biochemical characteristics.

(iv) Individuals with similar phenotype may be homozygous [RR] or heterozygous [Rr]. Hence they may or may not breed true.

(v) With complete dominance, phenotypic classes will always be lesser than genotypic classes.

(vi) For single gene difference with complete dominance, the observed phenotypic ratio in F2 would be 3:1.

(vii) Phenotype is the result of interaction between genotype and environment. Hence the expression of phenotype is influenced by environmental conditions.

(viii) A phenotype is governed by specific genes present at the molecular level [genotype].

(ix) The phenotype will change with change in genotype *i.e.* change in DNA.

(x) A phenotype is influenced by environmental factors.

Comparison of Genotype and Phenotype

Genotype and phenotypes differ in several aspects as presented in the Table 6.1.

TABLE 6-1: Comparison of Genotype and Phenotypes

Sl.No.	*Particulars*	*Genotype*	*Phenotype*
1	Determined by	Molecular structure	Morphological traits
2	Segregation classes	Higher than phenotypic classes	Lesser than genotypic classes
3	Alteration is possible by Mutation and hybridization	Possible	Possible.
4	Environmental effect	Little	High
5	Genes contained	Several hundreds	Several hundreds
6	Stability	High	Low to moderate
7	Outward differences	Absent	Present
8	Controlled by	Genes	G x E Interaction
9	Term coined by	Johannsen, 1909	Johannsen, 1909.

Genotyping

Genotyping refers to molecular characterization of an individual. In other words, it is the process of measuring genetic variations (polymorphism) in the DNA sample of a genotype. Main features of genotyping are presented below:

(i) Genotyping measures molecular features [DNA polymorphism] of an individual.

(ii) It is applicable to both animals as well as plants.

(iii) It helps in identification of useful alleles (disease resistance) in a genotype.

(iv) It helps in locating useful mutant alleles in a genotype.

(v) It is also known as DNA fingerprinting.

(vi) Various DNA markers such as AFLP, RFLP, RAPD, SNP, SSR, *etc.* are used for genotyping.

(vii) It also helps in allele mining *i.e.* discovery of new alleles for new uses.

(viii) Genotyping can be done for single locus as well as for multiple loci.

(ix) For genotyping, DNA can be isolated from any plant part such as root, stem, leaves, flowers and seeds.

Advantages

(i) Genotyping technique is simple and faster.

(ii) It is lesser expensive than phenotyping.

(iii) It can detect recessive mutant alleles which do not express in phenotype.

(iv) It can detect harmful alleles such as disease causing mutant alleles.

(v) It is useful in marker assisted selection and characterization of gene pool.

(vi) It is useful in the discovery of new useful alleles.

Limitations

(i) Huge data is generated by analysis of genotyping. Handling of such data is difficult.

(ii) The is variation in the results of genotyping from laboratory to laboratory.

(iii) Skilled and technical person is required to get accurate and reliable results.

Applications

Genotyping has three important practical applications in the field of plant breeding, research and marketing as discussed below:

(i) Plant Breeding: In plant breeding important applications include characterization of germplasm, varietal identification and marker assisted selection.

(ii) Research: Important research applications include identification of the parents of a hybrid and tracing genetic evolution of different crops. Microsatellite and SNP markers are used for this purpose.

(iii) Marketing: Important marketing applications include protection of plant breeders' rights and patents, quality control in production and processing of food and other products derived from plants.

Phenotyping

Phenotyping refers to characterization of morphological features of an individual. In other words, it measures morphological or visible variation in an individual. Main features of phenotyping are listed below:

(i) It measures morphological features of an individual.

(ii) It determines difference between mutant and wild type alleles.

(iii) It correlate the morphological features of mutant alleles with its genetic makeup.

(iv) Various highly heritable features are used for phenotyping.

(v) Phenotyping is carried out with germplasm and other breeding material including wild species and wild relatives.

Comparison of Genotyping and Phenotyping

A brief comparison of genotyping and phenotyping is presented in the Table 6.2.

TABLE 6-2: Comparison of Genotyping and Phenotyping

Sl.No.	*Particulars*	*Genotyping*	*Phenotyping*
1	Characterization done	Molecular	Morphological
2	Basis of analysis	DNA markers	Visual traits
3	Helps in identification of	Desirable alleles	Desirable traits
4	Skill Required	High	Average to moderate
5	Cost involved	High	Low

Advantages

(i) Helps in identification of resistant or susceptible to a particular disease.

(ii) Detection of the effect of mutation is easy.

(iii) Interpretation of results is easy.

(iv) There is less inter-laboratory variation in the results.

(v) It is less expensive, because it does not require sophisticated laboratory.

Limitations

(i) It is possible when plant are fully grown.

(ii) We have to wait for results till maturity is attained.

(iii) Phenotype is not always an indication of good genotype.

(iv) Phenotype test sometimes fail to detect effect of micro-mutations.

SUMMARY

Knowledge of genotype and phenotype is essential in genetics and plant breeding. Hence, it is necessary to define these terms before discussing genotyping and phenotyping. Terms genotype and phenotype were first used by Johannsen, a Danish Biologist, in 1909 working with common bean (*Phaseolus vulgaris*).

The genetic constitution or make up of an individual is known as genotype. In other words, genotype is the structure of DNA which determines the expression of a trait. Phenotype refers to external appearance of an individual. In other words, it refers to visual characteristics of an individual such as colour, height, sex, *etc.* For example, purple and white flower colours are two different phenotypes. Important features of genotype and phenotype have been highlighted.

Genotyping refers to molecular characterization of an individual. In other words, it is the process of measuring genetic variations (polymorphism) in the DNA sample of a genotype. Phenotyping refers to characterization of morphological features of an individual. In other words, it measures morphological or visible variation in an individual. Important features, advantages and limitations of genotyping and phenotyping are highlighted.

Various applications of genotyping in the field of plant breeding, research and marketing are discussed.

QUESTIONS

1. **What is genotyping? Describe briefly its advantages and limitations.**
2. **Describe various applications of genotyping.**
3. **What is phenotyping? Describe briefly its advantages and limitations.**
4. **Give a comparison of genotype and phenotype in tabular form.**
5. **Explain differences between genotyping and phenotyping in tabular form.**
6. **Explain briefly advantages and limitations of genotyping and phenotyping.**
7. **Write short notes on the following:**

   (i) Genotype (ii) Phenotype

   (iii) Genotyping (iv) Phenotyping.

Chapter 7

Quantitative Trait Locus (QTL) Mapping

INTRODUCTION

The quantitative trait locus (QTL) may be defined as a region or locus on the chromosome containing alleles that differently affect expression of a quantitative trait. In other words, QTL is a region of DNA which is associated with the expression of a quantitative trait. The main points related to QTL are given below.

1. The QTL is a polymorphic site (locus) on the chromosome which controls expression of a quantitative trait.
2. The presence of QTL is inferred from genetic mapping.
3. A set of genes collectively control a quantitative trait.
4. Ten or more QTL can influence a single trait.
5. The variation in a quantitative trait depends on the number of alleles governing such trait.
6. Alleles controlling expression of a quantitative trait may be present on one chromosome or different chromosomes in the genome.

QTL Mapping

Gene mapping is a technique which is used to identify genes responsible for expression of specific traits. The QTL mapping refers to location genes that control expression of quantitative traits. There are some requirements of QTL mapping such as (i) sophisticated laboratory, (ii) trained man power, (iii) mapping populations, (iv) molecular markers, and (v) mapping techniques. These are briefly discussed as follows:

1. **Sophisticated Laboratory:** A well equipped laboratory is required for performing QTL mapping. Several research laboratories are engaged in QTL mapping in different research Institutes the world over.

2. **Trained Manpower:** Trained manpower is essential to perform QTL mapping work. Now sufficient manpower is available for this purpose in different countries. Scientists are trained in molecular techniques under human resource development in advanced countries.
3. **Mapping Populations:** For QTL mapping a mapping population is required. Such population is derived from a cross between two diverse parents differing for the character of interest. The breeding materials which are commonly used as mapping populations include (i) F2 population, (ii) backcross populations, (iii) doubled haploid lines, (iv) recombinant inbred lines-RIL, (v) near isogenic lines-NIL, and (vi) bulk segregant analysis (BSA). All these populations have some merits and demerits.
4. **Molecular Markers:** Molecular markers or DNA markers are used for QTL analysis. A variety of molecular markers are now available for QTL analysis. The important molecular markers include RFLP, AFLP, RAPD, CAPS, SSR and SNP.
5. **Mapping Techniques:** Several mapping techniques are available for QTL mapping in crop plants. Important QTL mapping techniques include (i) single marker regression, (ii) simple interval mapping, (iii) composite interval mapping, (iv) multiple interval mapping, and (v) Bayesian interval mapping.

Methods of QTL Mapping

As stated above there are five methods of QTL mapping, *viz*.. (i) single marker regression, (ii) simple interval mapping, (iii) composite interval mapping, (iv) multiple interval mapping, and (v) Bayesian interval mapping. These are briefly discussed as follows:

1. **Single Marker Analysis:** It is the easiest and simplest method of QTL analysis. It is based on the idea that if there is an association between a marker genotype and trait value, it likely that a QTL is close to that marker locus. Thus it finds association between marker genotype and trait value. It involves regression analysis. It is used for quick scanning of entire genome to find out best possible QTLs. It is also used to identify missing or incorrectly formatted data. There are two main limitations of this method. Firstly, it under estimates QTL number and secondly, it cannot determine QTL position precisely.
2. **Interval Mapping:** It is an extension of single marker analysis. It overcomes the drawbacks of single marker analysis. Interval mapping provides a systematic way to scan the whole genome for evidence of QTL. It provides accurate results about the number and position of QTL. Single marker analysis uses only one marker for QTL mapping, whereas interval mapping uses two observable flanking markers to construct an interval within which to search for QTL. It permits detection of QTL in each interval lying between any two flanking markers. It can detect QTL

which cannot be detected single marker analysis. It is also known as simple interval mapping.

TABLE 7-1: Comparison of Interval Mapping and Multiple Interval Mapping

Sl.No.	*Particulars*	*Interval Mapping*	*Multiple Interval Mapping*
1	Marker used	Single	Several
2	Precision	Moderate	Very high
3	QTL number	Under estimated	Proper estimation
4	QTL position	Rough idea	Accurate
5	Technique	Simple	Difficult

3. **Composite Interval Mapping:** It is a modification of simple interval mapping. In this method, genetic markers are incorporated as co-factors in frame work of simple interval mapping. It greatly increases the power for detecting QTL and estimating QTL positions. Advantages of this method are as follows:
   (i) It has more power and precision than interval mapping in detecting number and position of QTLs.
   (ii) It eliminates the effects of other QTLs. Thus, it provides précised results by removing the effects of other QTLs linked to the locus under testing.

   There are some drawbacks of this method. It is highly dependent on the background information. Composite Interval Mapping (CIM) with too many background is difficult.
4. **Multiple Interval mapping:** It is modification of simple interval mapping. It utilizes multiple marker intervals simultaneously to fit multiple putative QTL directly in the model for mapping QTL. It provides information about number and position of QTL in the genome. It also determines interaction of significant QTLs and their contribution to the genetic variance. It is based on Cockerham's model for interpreting genetic parameters. Moreover, it is base on maximum likelihood for estimating genetic parameters.
5. **Bayesian Interval Mapping:** This technique was developed by Satagopon *et al.*, in 1996. It provides a model for QTL mapping. It provides information about number and position of QTL and their effects. It also provides information about position estimates of multiple QTL in an interval. It can estimate QTL effects and position separately.

Advantages

QTL mapping has several useful applications in plant breeding. In other words, it has many advantages over conventional methods of plant breeding. Important advantages of QTL mapping are briefly discussed as follows:

1. QTL mapping provides a systematic and more reliable approach for dissecting polygenic variation.
2. It is useful in determining or identifying useful QTLs in different breeding lines. Thus, lines with valuable QTLs can be selected for use in the hybridization programs.
3. The knowledge of QTL maps is very much useful in marker assisted selection. MAS can be used for genetic enhancement of elite breeding material. Valuable single genes can be introgressed from exotic donors or wild relatives an land races.
4. Marker assisted breeding (MAB) permits faster recovery of recipient parent genome than conventional recurrent backcrossing.
5. The application of markers for tagging of simply inherited traits such as disease resistance would be rewarding.
6. Marker assisted breeding (MAB) is usually superior to conventional (phenotype based) approach for improving traits with low heritability such as yield.

Limitations

There are some limitations of QTL mapping which are briefly presented as follows:

1. The QTL mapping is an expensive technique, because it requires well equipped laboratory, costly chemicals and glassware and skilled manpower.
2. The QTL mapping is difficult because alleles governing quantitative traits are distributed in the entire genome.
3. QTL mapping requires large number of gene markers which is a cumbersome work.
4. It requires information about easily recognizable single gene loci, which may not be available sometimes.
5. The QTL detected by any technique is not a true gene. It only indicates the genome region that most likely contain gene for the trait under study.
6. In the expression of quantitative traits, polygenic control, epistasis and G x E interaction are involved. Existing QTL detection techniques do not seem to have the required precision to deal with these problems.

FUTURE THRUSTS

Some of the future thrust areas of QTL mapping are indicated here under which should be given due importance.

1. There is need to further refine QTL mapping techniques to provide précised results for quantitative traits involving epistasis and high amount of G x E interactions.

2. The QTL with major effect can be easily manipulated by conventional breeding techniques. Hence marker technology should be used for mapping of those QTLs that show relatively minor effects (Stubber, *et al.*, 1999).
3. The present methods of QTL mapping are based on the assumption that epistasis is absent which is not true. The epistasis should be taken care of in QTL detection model.
4. The QTL mapping team should be interdisciplinary involving scientists from biochemistry, physiology, genetics and plant breeding. In facts, plant breeders are actual users of the results obtained from molecular mapping technology. QTLs have been mapped in several crop plants such as rice, maize, tomato, Arabidopsis, *etc.*

SUMMARY

The QTL may be defined as a region or locus on the chromosome containing alleles that differently affect expression of a quantitative trait. In other words, QTL is a region of DNA which is associated with the expression of a quantitative trait.

The QTL mapping refers to location of genes that control expression of quantitative traits. There are some requirements of QTL mapping such as (i) sophisticated laboratory, (ii) trained man power, (iii) mapping populations, (iv) molecular markers, and (v) mapping techniques.

As stated above there are five methods of QTL mapping, *viz.*. (i) single marker regression, (ii) simple interval mapping, (iii) composite interval mapping, (iv) multiple interval mapping, and (v) Bayesian interval mapping. These are briefly discussed.

QTL mapping has several useful applications in plant breeding. In other words, it has many advantages over conventional methods of plant breeding. Important advantages of QTL mapping have been discussed. Limitations and future thrusts of QTL mapping have been explained.

QUESTIONS

1. **What is QTL mapping? What are basic requirements of QTL mapping?**
2. **Describe briefly various methods of QTL mapping.**
3. **What is interval mapping? Describe its merits and demerits.**
4. **What is composite interval mapping? Describe its merits and demerits.**
5. **What is multiple interval mapping? Describe its merits and demerits.**
6. **Describe briefly Bayesian interval mapping. Describe its merits and demerits.**
7. **Explain briefly advantages and limitations of QTL mapping.**

8. **Write short notes on the following**
   (i) Composite interval mapping
   (ii) Multiple interval mapping

9. **Define the following terms**
   (i) QTL
   (ii) Gene mapping
   (iii) QTL Mapping
   (iv) Interval mapping

Chapter 8

DNA Fingerprinting

INTRODUCTION

DNA fingerprinting technique was developed by Alec Jeffrey in 1984 for identification of individuals based on DNA sequences. This is a technique by which an individual can be identified at molecular level. It is a quick way to analyze and compare the DNA sequences of living organisms. DNA fingerprinting can be done with DNA of plants as well as animals.

(i) In plants, DNA finger printing can be done by isolating DNA from root, stem, leaves, flower and seeds.

(ii) In animals, DNA finger printing can be done by isolating DNA from blood, muscle, bones, teeth, semen, hair, *etc.*

(iii) The DNA profile of an individual is unique. It can never be identical even in closely related individuals except for identical (mono-zygous) twins.

(iv) In DNA fingerprinting, laboratory DNA samples for testing are received either by the honorable court or through the police officer. The DNA test results are considered valid in the court of law.

Methods of DNA Fingerprinting

The DNA fingerprinting methods are also known as DNA testing techniques. Different DNA markers techniques are used for DNA fingerprinting. The most commonly used techniques are:

1. Restriction fragment length polymorphism [RFLP]
2. Amplified fragment length polymorphism [AFLP]
3. Random amplified polymorphic DNA [RAPD]
4. Simple sequence repeats or micro-satellites [SSR] and
5. Single nucleotide sequence [SNP].

Each technique has its advantages and disadvantages. RFLP technique is based on Southern blotting. Now, PCR based techniques are becoming more popular because of their several advantages over Southern blotting. Moreover, DNA typing can be done using partially degraded DNA.

Types of DNA Fingerprinting

There two types of DNA fingerprinting techniques which are commonly used, *viz..*, single locus DNA finger printing and multi-locus DNA fingerprinting. These are explained as follows:

(i) **Single Locus DNA Fingerprinting**: This technique is useful when DNA is degraded and for mixed samples. Polymorphism at a single locus is analyzed usually through use of specific DNA probes or specific PCR primers.

(ii) **Multi-locus DNA Fingerprinting:** This technique provides more information per sample than single locus fingerprinting. Polymorphism at multi-locus is simultaneously identified. This can be performed by application of mixture of single locus probes or application of single locus that identifies multiple similar sequence polymorphisms. The second case detects unidentified fragments of DNA. Thus the result is a DNA phenotype rather than a genotype.

Steps Involved

DNA fingerprint is a pattern generated during analysis of DNA in the laboratory. It is unique to an individual. The DNA fingerprint is the same for every cell, tissue or organ of an organism. It cannot be altered by any known treatment. DNA fingerprinting is a laboratory procedure which consists of following six steps, *viz..* (i) isolation of DNA, (ii) cutting, sizing and storing of DNA, (iii) transfer of DNA to nylon, (iv) probing, (v) washing of probe and (vi) DNA fingerprints. These are briefly discussed as follows:

(i) **Isolation of DNA:** The DNA is isolated from the cells of plants or animals. Very small amount of DNA is required for testing. For example, the amount of DNA found I one drop of blood or one square centimeter area of leaf tissue is usually sufficient.

(ii) **Cutting, sizing and sorting of DNA:** Special enzymes called restriction enzymes are used to cut the DNA at specific places. For example, an enzyme called ECOR 1 found in bacteria will cut DNA when the sequence GAATTC occurs. Even one base change in the restriction enzyme cutting site will prevent enzyme from cutting the DNA. Thus a very small difference in the DNA of two individuals can result in cut pieces of DNA being of different size and in turn can cause the DNA fingerprints of these two individuals to differ.

The DNA pieces are sorted according to size by a sieving technique called electrophoresis through a gel made from seaweed agarose. Electrophoresis is a laboratory procedure in which DNA pieces are size sorted by travelling through an agarose gel in an electric field.

(iii) Transfer of DNA to Nylon: The DNA pieces are transferred to a nylon sheet placing the agarose gel and nylon sheet next to each other over night.

ISOLATION OF DNA

↓

CUTTINIG and SORTING OF DNA

↓

TRANSFER OF DNA TO NYLON SHEET

↓

PROBING

↓

WASHING

↓

DNA FINGERPRINTS

FIGURE 8-1: Steps Involved in DNA Fingerprinting.

(iv) Probing: Probes are very small pieces of DNA used to find specific sequence of letters in a very long piece of DNA from a chromosome. The DNA finger print is generated by adding tagged probes to the nylon sheet. Each probe typically sticks in only one or two specific places whenever the sequences match [A with T and G with C]. The tag allows detection of the probe.

(v) Washing: The excess probe material is washed away leaving the unique DNA band pattern.

(vi) DNA Fingerprints: The final DNA fingerprints are built from several different probes. The radioactive DNA pattern is transferred to X-ray film direct exposure. When developed, the resultant visible pattern is the DNA fingerprint which looks like the bar codes used at the grocery counter.

Advantages

DNA Fingerprinting has several advantages. Some important advantages of DNA fingerprinting are presented here as under.

1. **Unique Feature:** Every individual has unique DNA fingerprints which remain same throughout life, even after death.
2. **No alteration:** DNA fingerprints cannot be altered by any means. Moreover, DNA fingerprints cannot be duplicated. DNA fingerprints cannot be erased or modified by any means.
3. **No hiding:** A person cannot hide his DNA fingerprints.
4. **Reliable Technique:** DNA fingerprinting is a reliable technique for identification of individuals. Reliable information can be obtained even from degraded (years old) samples of DNA.
5. **Testing from any Sample:** DNA testing can be performed even with the samples that have been exposed to detergents, acids, bases, salts, gasoline and bacterial contamination.
6. **Requires Small sample:** For DNA fingerprinting, very small sample is required. For example, DNA testing can be performed from the roots of a few hairs.

Drawbacks

There are several advantages of DNA fingerprinting technique as discussed above. However, there are some drawbacks of this technique which are listed as follows:

1. The DNA fingerprinting is not 100 per cent reliable. Even 1 case in 20 billions may lead to wrong decision.
2. DNA testing from persons of blood transfusion may give misleading results.
3. The contamination of blood samples or wrong labels may give misleading picture.
4. Error in hybridization and probing process will give wrong results.
5. Error by technician may not be 100 per cent ruled out.
6. The testing cost is very high which cannot be afforded by poor people.

There are some differences between DNA fingerprinting and conventional fingerprints which are presented in the Table 8.1.

TABLE 8-1: Differences between DNA Fingerprints and Conventional Fingerprints

Sl.No.	*Particulars*	*DNA Fingerprints*	*Fingerprints*
1	Test conducted with	DNA	Finger tips.
2	Testing	Expensive	Cheap
3	Alteration	Not possible	Possible by surgery
4	Duplication	Not possible	Possible
5	Applications	Wide	Limited

Applications in Crop Improvement

DNA Fingerprinting or DNA testing has several practical applications in crop improvement, In crop improvement, DNA testing is useful in four various ways, *viz..* (i) assessment of diversity, (ii) identification of transgenic plants, (iii) marker assisted selection, (iv) protection of legal rights, (v) crop evolution, (vi) parentage detection, (vii) disease detection, (viii) predicting hybrid performance, and (ix) gene mapping. These are briefly discussed as follows:

1. **Assessment of Diversity:** Molecular markers can be used for assessment of genetic diversity in cultivars, germplasm collections, and advanced breeding material. This information can be used for germplasm characterization and developing varietal information system and plant genetic resources information system. This will also help in patenting plant material with special characters. The knowledge of genetic diversity will help in selection of parental lines for development of high yielding hybrids and cultivars. Crosses between distantly related parents can be made to obtain highly heterotic combinations.
2. **Identification of Transgenic Plants:** Transgenic refers to the organism developed by the technique of genetic engineering bypassing sexual process. DNA fingerprinting is useful in the identification of transgenic plants.
3. **Marker Assisted Selection**: The DNA fingerprinting is used in marker assisted selection. In marker assisted selection, plants with a piece of DNA marker associated with a desirable trait are selected. MAS is useful in early selection of desirable plants without growing them to maturity. Thus DNA fingerprinting speeds up the selection process and reduce the cost considerably. The DNA fingerprinting can distinguish homozygous individuals [AABBCC] from heterozygous ones [AaBbCc] and can make the selection process more precise. It is not possible to distinguish between homozygotes and heterozygotes on the basis of phenotypes.
4. **Protection of Legal Rights:** The DNA fingerprinting is used to legally protect new varieties of plants and breeds of animals, whether they were developed by genetic engineering, tissue culture or conventional methods. In other words, DNA fingerprinting is used for protection of plant breeders' rights and patents.
5. **Crop Evolution:** DNA markers are useful in the study of evolution. Molecular markers help in tracing the genetic origin of crop plants. It will help in identification of the wild species involved in the genetic evolution of different crop plants. Molecular markers are also useful in discovery or identification of new alleles in the germplasm or wild species of crop plants.
6. **Parentage Detection:** DNA testing of hybrid and its parental lines will provide information about the inbred lines involved in the hybridization.

The DNA of hybrid has some resemblances with the DNA of female and male lines involved in the hybridization.

7. **Disease Detection:** DNA testing of can be used as reliable technique in detecting the presence of a pathogen in the plants even if it is present in the latent stage. Such plants can be discarded from use in the hybridization programs.
8. **Predicting Hybrid performance:** Based on the assessment of genetic diversity of inbred lines or parental lines, the performance of hybrids can be easily predicted. The DNA fingerprints of inbred lines can be used in prediction the performance of hybrid resulting from the cross of such lines. The DNA fingerprints will provide information about the genetic diversity of inbred lines which in turn can be used in prediction expected performance of their hybrid progeny.
9. **Gene Mapping:** Gene mapping is a method which is used to identify the genes responsible for expression of specific traits. DNA markers are useful in determining the location of genes that control important traits. The DNA fingerprinting is used to identify DNA markers which are inherited along with important traits.

SUMMARY

A nucleic acid that carries the genetic information in the cell and is capable of self-replication and RNA synthesis is referred to as DNA [Deoxyribose Nucleic Acid]. The scientific name for DNA is deoxyribonucleic acid. The main features of DNA in terms of location, structure, shape, replication, types, functions and composition have been briefly discussed.

DNA fingerprinting is a technique by which an individual can be identified at molecular level. It is a quick way to analyze and compare the DNA sequences of living organisms. DNA fingerprinting can be done with DNA of plants as well as animals. DNA Fingerprinting has several advantages. Important advantages and limitations of DNA fingerprinting have been discussed.

DNA Fingerprinting or DNA testing has several practical applications in crop improvement, In crop improvement, DNA testing is useful in four various ways, *viz*.. (i) assessment of diversity, (ii) identification of transgenic plants, (iii) marker assisted selection, (iv) protection of legal rights, (v) crop evolution, (vi) parentage detection, (vii) disease detection, (viii) predicting hybrid performance, and (ix) gene mapping. These are briefly discussed as follows:

QUESTIONS

1. **What are various steps involved in DNA fingerprinting?**
2. **What is DNA fingerprinting? Describe its advantages and limitations.**
3. **Discuss briefly various applications of DNA fingerprinting in crop improvement.**

4. **Explain applications of DNA fingerprinting in crop improvement in relation to (i) assessment of diversity and (ii) identification of transgenic plants.**
5. **Describe briefly applications of DNA fingerprinting in crop improvement on the basis of (i) parentage detection and (ii) disease detection.**
6. **Discuss briefly applications of DNA fingerprinting in crop improvement in relation to (i) marker assisted selection and (ii) protection of legal rights.**
7. **Explain briefly differences between DNA fingerprinting and conventional finger prints.**
8. **Write short notes on the following:**

   (i) DNA fingerprinting (ii) Conventional finger prints
9. **Describe briefly main features of DNA fingerprinting.**

Chapter 9

Methods of Foreign Gene Transfer

INTRODUCTION

All conventional plant breeding procedures result in genetic changes. All offspring are therefore technically genetically modified relative to their parents. In crop plants, artificial transfer of some desirable genes (traits) from one genotype to another is referred to as gene transfer.

Types of Gene Transfer

Gene transfer is of two types, *viz*.. horizontal gene transfer (HGT), and vertical gene transfer. Horizontal and vertical gene transfers are different phenomena and have different implications for the safety of genetically modified crops.

1. Horizontal gene Transfer

Horizontal gene transfer refers to any process in which an organism incorporates genetic material from another organism without being the offspring of that organism. It is also known as Lateral gene transfer (LGT). The important points related to horizontal gene transfer are discussed below.

1. **Importance**: Horizontal gene transfer is a highly significant phenomenon, and amongst single-celled organisms perhaps the dominant form of genetic transfer. Artificial horizontal gene transfer is a form of genetic engineering.
2. **Sexual Union**: In horizontal gene transfer, sexual reproduction is not involved. In other words, in horizontal gene transfer sexual process is bypassed.
3. **Occurrence**: Horizontal gene transfer achieved by human efforts. It seldom occurs under natural conditions.
4. **Overcomes Crossing Barriers**: Horizontal gene transfer overcome the boundaries of gene transfer between species. Now it is possible to transfer

a gene from a soil bacterium to a plant. For example, a freezing resistant gene has been transferred from fish to cultivated tomatoes. Similarly, ovalbumin gene of chicken has been transferred to alfalfa for improving the protein quality. Lot more such examples can be cited.

5. **Technique Used:** Horizontal gene transfer is achieved by the technique of biotechnology or genetic engineering.
6. **Methods of Gene Transfer**: There are four methods of horizontal gene transfer, *viz.*. agro-bacterium based gene transfer, particle gun, electro-poration and direct DNA up take. Detailed account of these methods is beyond the scope of present discussion.
7. **Application:** Horizontal gene transfer can be used for the genetic improvement of both self-pollinated and cross pollinated species. It can be used for genetic improvement of both seed propagated and vegetatively propagated species.
8. **Bio-safety Measures:** The safety assessment of transgenic plants includes assessing both the possibility of out-crossing to close relatives and evaluating the potential ecological consequences if it were to occur. If crossing takes place between a transgenic and a non-transgenic crop, the latter could acquire a novel gene. This is possible when a transgenic cultivated plant is likely to find a related wild plant as an out-crossing partner.
9. **End product**: The end product of a horizontal gene transfer is a transgenic plant or genetically engineered plant.
10. **Achievements:** Horizontal gene transfer has been used in many crop plants such as Rapeseed, Potato, Tobacco, Tomato, maize, Linseed, Soybean, Cotton, Sugar beet, Lucerne and many other crops. All these crops have been improved for disease resistance, insect resistance, herbicide resistance and quality through horizontal gene transfer. Horizontal gene transfer differs from vertical gene transfer several aspects (Table 9-1).

2. Vertical Gene Transfer

When an organism receives genetic material from its ancestor, *e.g.* its parent or a species from which it evolved it is known as *vertical transfer* gene transfer. Two plants crossing sexually and passing their genes on to following generations is an example of vertical gene transfer (usually called "crossing"). Gene transfer via pollen between plants of the same or related species takes place in the nature. Thus, the transfer of disease and pest resistance from cultivated plants to related wild species and vice versa has always taken place, regardless of how the resistance genes were acquired in the first place. The important points related to vertical gene transfer are discussed below.

1. **Importance**: Most thinking in plant breeding has focused upon vertical transfer. Vertical gene transfer is widely used for genetic improvement of crop plants. Artificial vertical gene transfer is a form of out-breeding.

TABLE 9-1: Differences between Horizontal Gene Transfer and Vertical Gene Transfer

Sl.No.	*Particulars*	*Horizontal Gene Transfer*	*Vertical Gene Transfer*
1	Used in	Single celled organisms	Used in multi-cellular organisms
2	Sexual process	Not Involved	Involved
3	Occurrence in nature	Seldom	Common
4	Crossing barriers	Not present	Present
5	Technique used	Genetic engineering	Conventional breeding methods
6	Methods used	Agrobacterium based method, gene gun micro injection direct DNA uptake	Pedigree, bulk breeding, single seed descent, back cross breeding.
7	Applications	In all crops	In all crops
8	Bio-safety measures	Required	Not required
9	End product	Transgenic plants	Non transgenic plants
10	Bioethical measures	Required	Not required
11	Evolution of super weeds	Possible	Not possible
12	Organic farming of end product	Not permitted	Permitted
13	Accuracy of method	Very high	Moderate to high
14	Time required for release of new variety	4-5 years	10-12 years
15	Technical skill required	Very high	Moderate
16	Laboratory required	Sophiscated	Ordinary
17	Initial cost of infra structure	Very high	Low
18	Direct single gene transfer	Possible	Not possible
19	Transgene	Involved	Not involved

2. **Sexual Union**: In vertical gene transfer, sexual process [union of male and female gametes] is involved. In other words, vertical gene transfer is achieved by artificial crossing between two different genotypes.
3. **Occurrence**: Vertical gene transfer is common in nature. The vertical gene transfer takes place due to out crossing or cross pollination.
4. **Crossing Barriers**: In vertical gene transfer there are crossing barriers. It is not possible to transfer a gene from an unrelated species. Moreover, it is difficult to transfer gene from a distantly related species or genera. Such distant cross are often associated with cross incompatibility, hybrid sterility, hybrid in-viability and hybrid break down.
5. **Technique Used:** Vertical gene transfer is achieved by conventional breeding techniques.
6. **Methods of Gene Transfer**: Vertical gene transfer is achieved by conventional breeding methods, *viz..* pedigree method, bulk breeding method, back cross method, single seed descent, *etc.*

7. **Application:** Vertical gene transfer is widely used for the genetic improvement of self-pollinated species, cross pollinated species and asexually propagated species.
8. **Bio-safety Measures:** Bio-safety measures apply only to transgenic plants. Those plants which arise spontaneously or by conventional cross, would not be subjected to such regulatory procedures.
9. **End product**: The end product of a vertical gene transfer is a non-transgenic plant.
10. **Achievements:** Vertical gene transfer has been widely used in all the crop plants for genetic improvement of various characters such as disease resistance, insect resistance, herbicide resistance and quality.

In future, horizontal gene transfer is expected to achieve tangible results in three main areas, *viz*.. development of resistant cultivars, improvement in quality and development of new useful products.

Steps Involved

Horizontal gene transfer is gaining increasing importance these days. Transgenic plants are developed though horizontal gene-transfer. The horizontal gene transfer is achieved through genetic engineering. The important steps in the development of transgenic plants include identification of the gene of interest, its isolation, cloning and transfer to the host plant.

Methods of Foreign Transfer

There are four important methods of foreign gene transfer. These are; plasmid method, particle bombardment, direct DNA uptake, and micro-injection. These methods are also known as systems of DNA delivery for genetic transformation. A brief description of these methods of foreign transfer is presented as follows:

1. Plasmid Method

Plasmids are extra chromosomal elements which are found in the bacterial cell. They are able to replicate independently of chromosomal DNA and are not essential for normal growth and function of a bacterium. Plasmids are used as cloning vectors in the genetic engineering. The soil borne bacterium *Agrobacterium tumefaciens* is used for development of transgenic plants. The agrobacterium mediated gene transfer is also known as biological method of foreign transfer. It consists of the following four main steps.

(i) Gene Cloning

Gene cloning is a technique of genetic engineering which is used to make multiple identical copies of a gene. Plasmids of Agrobacterium contain tumor producing genes [Ti genes or oncogenes]. The foreign gene which has to be used for genetic transformation is incorporated in the plasmid for cloning. Thus several identical copies of the trans gene are produced by self-replication of plasmids.

(ii) Genetic Transformation

The process of transfer, integration and expression of transgene or foreign gene in the host cells is known as genetic transformation. For genetic transformation, bacteria with cloned gene are mixed [in millions] in the cell culture or protoplast culture of the host plant for a day or two. Out of several thousands of cells or protoplasts in the culture medium, only a few are transformed. In other words, plasmid DNA [foreign DNA] can enter, integrate and express in few cells or protoplasts.

(iii) Identification of Transformed Cells

Several methods are available for detecting the presence of transformed cells. Such methods include selectable marker, PCR, ELISA, southern, northern and western blot techniques. A selectable marker is widely used for this purpose. The selectable marker is attached to the target gene for easy identification. Generally an antibiotic resistant gene is used as selectable marker to confirm whether the gene of interest has been inserted in the host plant or not. The Kanamycin resistant gene is used for identification of transformed cells. The lethal concentration of Kanamycin is put in the plasmid mixed cell suspension or protoplast culture. The non-transformed cells will die and only transformed cells will survive. Thus transformed cells are detected.

(iv) Regeneration of Transformed Cells

The transformed cells are transferred in to culture medium for transformation in to whole plant. The regeneration of cell or protoplasts in to whole plant is the basic requirement of this method of foreign gene transfer.

Advantages

There are two main advantages of Agrobacterium mediated method of DNA transfer. As given below.

(i) This method has some control over the copy number and site of integration of trans gene which is not possible in particle bombardment method.

(ii) This is a cheaper method of genetic transformation.

Limitations

There are three main limitations of Agrobacterium mediated method of genetic transformation as given below.

(i) Host Specificity: This method limited host range. It cannot infest monocotyledonous crops especially cereals.

(ii) Somaclonal variation: Lots of variation is induced in the tissue culture which poses problem in the identification of transformed cells.

(iii) Slow regeneration: Regeneration of cells or protoplasts into whole plant takes more time than regeneration from meristems and organ cultures.

2. Particle Bombardment Method

Particle bombardment method is also known as biolistic micro-projectile bombardment, particle acceleration and micro-ballistics. In this method, foreign DNA is delivered into host plant cells using high velocity metal particles through gene gun. Main features of this method are given as follows:

(i) In this method regenerable tissues such as meristems, immature embryos and embryogenic callus are used for bombardment.

(ii) The gold or tungsten particle [1-1.5 milli micron size] are used for bombardment.

(iii) The gold or tungsten particle coated with gene of interest are bombarded in the tissues with the help of gene gun.

(iv) The transformed tissues are identified by polymerase chain reaction [PCR] technique.

(v) The transformed tissues are regenerated in culture medium into whole plant, which further examined for stability of gene expression by ELISA, southern blot and northern blot techniques.

TABLE 9-2: Comparison of Plasmid and Gene Gun Methods of Foreign Transfer

Sl.No.	*Particulars*	*Plasmid Method*	*Gene Gun Method*
1	Other term used	Biological method of gene transfer	Particle bombardment, micro-ballistic and particle acceleration.
2	Basic requirement	Regenerable cells or protoplasts	Regenerable tissues or organs or cells.
3	Host specificity	Severe	No specificity
4	Effectiveness	More with dicots than monocots.	Equally effective with both groups.
5	Expenditure involved	Lower than gene gun method	Higher than plasmid method
6	Procedure	Difficult	Simple

Advantages

This method has some advantages over the Agrobacterium mediated method of DNA transfer. Advantages of this method are as follows:

(i) This method does not exhibit host specificity. Hence this method can be effectively used for development of transgenic plants in various plant species.

(ii) This method is technically simple than Agrobacterium mediated method of DNA transfer.

(iii) There is no need of isolating protoplasts. Various plant parts such as meristems, embryos or leaves can be bombarded and transformants can be regenerated from them.

Disadvantages

There are two main disadvantages of this method as stated below.

(i) There is no control over the copy number and site of integration of foreign gene.

(ii) The cost of equipment is high which prohibit use of this method by many researcher for DNA transfer.

3. Direct DNA Transfer

This method involves electroporation or chemical fusagens such as poly ethyl glycol [PEG] with calcium phosphate. This method is used with protoplasts. A suspension of protoplasts with desired DNA is prepared. Then high voltage current is applied though the protoplast DNA suspension. The electric current leads to formation of small temporary holes in the protoplast membrane though which the DNA can pass. After entry into the cell, the foreign DNA gets incorporated with the host genome, resulting in genetic transformation. Main points related to this method are given below:

(i) It is used with ptotoplast culture.

(ii) This is cheaper method of foreign transfer.

(iii) This is tedious than particle bombardment approach.

(iv) It exhibit host specificity.

4. Micro-Injection Method

In this method plasmid DNA is delivered in to host cells by microscopic needles, known as micro-injection. Main features of this method are presented below.

(i) It is used with protoplast culture. Regeneration from protoplast is difficult in many plant species.

(ii) It has no host specificity.

(iii) The regeneration from protoplasts is generally slow.

(iv) This is a cheaper method of foreign transfer.

Currently two DNA delivery systems, *viz*.. (i) Agrobacterium mediated gene transfer, and (ii) particle bombardment are widely used for genetic transformation in cotton. In India, six different events have been developed by these methods, which are being used for developing Bt. transgenic cotton hybrids. In India, 522 Bt. cotton hybrids have been developed for commercial cultivation till 2009.

SUMMARY

The gene transfer is of two types, *viz*.., vertical gene transfer and horizontal gene transfer. When an organism receives genetic material from its ancestor, *e.g.* its parent or a species from which it evolved, it is known as vertical gene transfer. The process in which an organism incorporates genetic material from another organism without being the offspring of that organism is called horizontal gene transfer (HGT) or lateral gene transfer (LGT).

Transgenic plants are developed though horizontal gene transfers. The horizontal gene transfer is achieved through genetic engineering. The important steps in the development of transgenic plants include identification of the gene of interest, its isolation, cloning and transfer to the host plant.

There are four important methods of foreign gene transfer. These are; plasmid method, particle-bombardment, direct DNA uptake, and micro-injection. All these methods have been amply discussed.

QUESTIONS

1. **Describe procedure, limitations and advantages of biological method of foreign transfer.**
2. **Discuss briefly procedure, limitations and advantages of particle method of foreign transfer.**
3. **Explain briefly procedure, limitations and advantages of micro-injection method of foreign transfer.**
4. **Describe briefly procedure, limitations and advantages of direct method of foreign transfer.**
5. **What are methods of gene transfer? Describe horizontal method of gene transfer. Give its merits and demerits.**
6. **Describe briefly Vertical method of gene transfer. Give its merits and demerits.**
7. **Explain briefly differences between horizontal and vertical methods of gene transfer.**

Molecular Markers

INTRODUCTION

Those characters which can be easily identified are called marker characters. Any genetic element (locus, allele, DNA sequence or chromosome feature) which can be readily detected by phenotype, cytological or molecular techniques, is referred to as marker. Markers related to variations in DNA fragments generated by restriction endonuclease enzymes are called DNA markers or genetic markers. Some other definitions of DNA markers are given below:

1. Any unique DNA sequence which can be used in DNA hybridization, PCR or restriction mapping experiments to identify that sequence is called DNA marker.
2. A gene or DNA sequence having a known location on a chromosome and associated with a particular gene or trait refers to DNA marker.

Types of Marker

Markers are of four types, *viz*.. (i) morphological, (ii) biochemical, (iii) cytological, and (iv) DNA markers. These are briefly discussed as follows:

1. Morphological Markers

In plant breeding, markers that are related to variation in shape, size, colour and surface of various plant parts are called morphological markers. Such markers refer to available gene loci that have obvious impact on morphology of plant. Genes that affect form, coloration, male sterility or resistance among others have been analyzed in many plant species. In rice, examples of this type of marker may include the presence or absence of awn, leaf sheath coloration, height, grain color, aroma *etc.* In well-characterized crops like maize, tomato, pea, barley or wheat, tens or even hundreds of such genes have been assigned to different chromosomes. There are several demerits of morphological markers as given below.

1. They generally express late into the development of an organism. Hence their detection is dependent on the development stage of the organism.
2. They usually exhibit dominance
3. Sometimes they exhibit deleterious effects
4. They exhibit pleiotropy
5. They exhibit epistasis
6. They exhibit less polymorphism
7. They are highly influenced by the environmental factors.

2. Biochemical Markers

Markers that are related to variation in proteins and amino acid banding pattern are known as biochemical markers. A gene encodes a protein that can be extracted and observed; for example, isozymes and storage proteins.

3. Cytological Markers

Markers that are related to variation in chromosome number, shape, size and banding pattern are referred to as cytological markers. In other words, it refers to the chromosomal banding produced by different stains; for example, G banding.

4. DNA Markers

A gene or other fragment of DNA whose location in the genome is known is called DNA marker. It is a unique (DNA sequence), occurring in proximity to the gene or locus of interest. It refers to any unique DNA sequence which can be used in DNA hybridization, PCR or restriction mapping experiments to identify that sequence. It can be identified by a range of molecular techniques such as RFLPs, RAPDs, AFLP, SNPs, SCARs, microsatellites *etc.* DNA markers are also known as molecular markers or genetic markers. To overcome problems associated with morphological markers, the DNA-based markers have been developed. Advantages of DNA markers are presented as follows:

1. They are highly polymorphic.
2. They have simple inheritance (often codominant).

TABLE 10-1: Comparison of Morphological Markers and DNA Markers

Sl.No.	*Particulars*	*Morphological Markers*	*DNA Markers*
1	Nature	Dominant	Co-dominant
2	Polymorphism	Low	High
3	Occurrence	Low	Abundant
4	Detection	Easy	Easy
5	Pleiotropic effect	High	Minimum
6	Epistasis	Present	Absent
7	Environmental effect	High	No effect

3. They abundantly occur throughout the genome.
4. They are easy and fast to detect.
5. They exhibit minimum pleiotropic effect.
6. Their detection is not dependent on the developmental stage of the organism.

Properties of an Ideal DNA Marker

An ideal DNA marker should have some properties or characteristics. Important properties of an ideal DNA marker are presented below:

(i) Markers should exhibit high level of polymorphism. In other words, there should be variability in the markers. It should demonstrate measurable differences in expression between trait types and/or gene of interest.

(ii) Marker should be co-dominant. It means, there should be absence of intra-locus interaction. It helps in identification of heterozygotes from homozygotes.

(iii) The marker should be multi-alleleic. It useful in getting more variability/ polymorphism for a character.

(iv) There should be absence of epistasis. It makes Identification of all phenotypes (homo- and heterozygotes) easy.

(v) The marker should be neutral. The substitution of alleles at the marker locus should not alter the phenotype of an individual. This property is found in almost all the DNA markers.

(vi) Markers should be insensitive to environment. This property is also found in almost all the DNA markers.

Types of DNA Markers

There are several types of DNA markers that are used in genetics and plant breeding for various purposes. Techniques for generating DNA markers have undergone rapid change in the last 30 years. The use of a specific marker technique is determined by the crop species, technical expertise, lab equipment, and research funding.

The most commonly-used types of DNA markers include restriction fragment length polymorphism (RFLP), random amplified polymorphic DNA (RAPD), inter-simple sequence repeat (ISSR), simple sequence repeats (SSR) and amplified length fragment polymorphism (AFLP. The list of commonly used DNA markers is presented below:

1. Restriction Fragment Length Polymorphism (RFLP)
2. Amplified Fragment Length Polymorphism (AFLP)
3. Random Amplified Polymorphic DNA (RAPD)
4. Cleaved Amplified Polymorphic Sequences (CAPS)

5. Simple Sequence Repeat (SSR) Length Polymorphism
6. Single Strand Conformational Polymorphism (SSCP)
7. Heteroduplex Analysis (HA)
8. Single Nucleotide Polymorphism (SNP)
9. Expressed Sequence Tags [EST]
10. Sequence Tagged Sites [STS]

1. Restriction Fragment Length Polymorphisms (RFLPs)

RFLPs refer to variations found within a species in the length of DNA fragments generated by specific endonuclease. RFLPs are first type of DNA markers developed to distinguish individuals at the DNA level. RFLP technique was developed before the discovery of Polymerase Chain Reaction [PCR]. The advantages, disadvantages and uses of this technique are presented below.

Advantages

RFLP technique has several advantages as given below.

(i) It is a cheaper and simple technique of DNA sequencing.
(ii) It does not require special instrumentation.
(iii) The majority of RFLP markers are co-dominant and highly locus specific.
(iv) These are powerful tools for comparative and synteny mapping.
(v) It is useful in developing other markers such as CAPS and INDEL.
(vi) Several samples can be screened simultaneously by this technique using different probes.
(vii) RFLP genotypes for single copy or low copy number genes can be easily scored and interpreted.

Disadvantages

(i) Developing sets of RFLP probes and markers is labour intensive.
(ii) This technique requires large amount of high quality DNA.
(iii) The multiplex ratio is low, typically one per gel.
(iv) The genotyping throughput is low.
(v) It involves use of radioactive chemicals.
(vi) RFLP finger prints for multi-gene families are often complex and difficult to score.
(vii) RFLP probes cannot be shared between laboratories.

Uses

(i) They can be used in determining paternity cases.
(ii) In criminal cases, they can be used in determining source of DNA sample.

(iii) They can be used to determine the disease status of an individual.

(iv) They are useful in gene mapping, germplasm characterization and marker assisted selection.

(v) They are useful in detection of pathogen in plants even if it is in latent stage.

2. Amplified Fragment Length Polymorphism (AFLP)

AFLPs are differences in restriction fragment lengths caused by SNPs or INDELs that create or abolish restriction endonuclease recognition sites. AFLP assays are performed by selectively amplifying a pool of restriction fragments using PCR.RFLP technique was originally known as selective restriction fragment amplification.

Advantages

1. Very high multiplex ratio and genotyping throughput.
2. Highly reproducible across laboratories.
3. No marker development work is needed; however, AFLP primer screening is often necessary to identify optimal primer specificities and combinations.
4. No special instrumentation is needed for performing AFLP assays; however, special instrumentation is needed for co-dominant scoring (see weaknesses below).
5. Start-up costs are moderately low.
6. AFLP assays can be performed using very small DNA samples (typically 0.2 to 2.5 μg per individual).
7. The technology can be applied to virtually any organism with minimal initial development.

Disadvantages

1. The maximum polymorphic information content for any bi-allelic marker is 0.5.
2. High quality DNA is needed to ensure complete restriction enzyme digestion.
3. DNA quality may or may not be a weakness depending on the species. Rapid methods for isolating DNA may not produce sufficiently clean template DNA for AFLP analysis.
4. Proprietary technology is needed to score heterozygotes and ++ homozygotes. Otherwise, AFLPs must be dominantly scored. Dominance may or may not be a weakness depending on the application.
5. The homology of a restriction fragment cannot be unequivocally ascertained across genotypes or mapping populations.

6. Developing locus specific markers from individual fragments can be difficult and does not seem to be widely done.
7. The switch to non-radioactive assays has not been rapid. Chemiluminescent AFLP fingerprinting methods have been developed and seem to work well (Lin *et al.*, 1997). The fingerprints produced by fluorescent AFLP assay methods are often difficult to interpret and score and thus do not seem to be widely used.
8. AFLP markers often densely cluster in centromeric regions in species with large genomes, *e.g.*, barley (*Hordeum vulgare* L.) (Qi *et al.*, 1998) and sunflower (*Helianthus annuus* L.) (Gedil *et al.*, 2000).

Uses

This technique has been widely used in the construction of genetic maps containing high densities of DNA marker. In plant breeding and genetics, AFLP markers are used in varietal identification, germplasm characterization, gene tagging and marker assisted selection.

3. Random Amplified Polymorphic Sequences (RAPDs)

RAPD refers to polymorphism found within a species in the randomly amplified DNA generated by restriction endonuclease enzyme. RAPDs are PCR based DNA markers. RAPD marker assays are performed using single DNA primer of arbitrary sequence. DNA polymorphisms produced by "rearrangements or deletions at or between oligonucleotide primer binding sites in the genome" (Williams *et al.*, 1990; Welsh and McClelland 1990).

Advantages

1. RAPD primers are readily available being universal.
2. They provide moderately high genotyping throughput.
3. This technique is simple PCR assay (no blotting and no radioactivity).
4. It does not require special equipment. Only PCR is needed.
5. The start-up cost is low.
6. RAPD marker assays can be performed using very small DNA samples (5 to 25 ng per sample).
7. RAPD primers are universal and can be commercially purchased.
8. RAPD markers can be easily shared between laboratories.
9. Locus-specific, co-dominant PCR-based markers can be developed from RAPD markers.
10. It provides more polymorphism than RFLPs.

Disadvantages

1. The detection of polymorphism is limited. The maximum polymorphic information content for any bi-allelic marker is 0.5.

2. This technique only detects dominant markers.
3. The reproducibility of RAPD assays across laboratories is often low.
4. The homology of fragments across genotypes cannot be ascertained without mapping.
5. It is not applicable in marker assisted breeding programme.

Uses

This technique can be used in various ways such as for varietal identification, DNA fingerprinting, gene tagging and construction of linkage maps. It can also be used to study phylogenetic relationship among species and sub-species and assessment of variability in breeding populations.

4. Cleaved Amplified Polymorphic Sequences (CAPS)

CAPS polymorphisms are differences in restriction fragment lengths caused by SNPs or INDELs that create or abolish restriction endonuclease recognition sites in PCR amplicons produced by locus-specific oligonucleotide primers

1. CAPS assays are performed by digesting locus-specific PCR amplicons with one or more restriction enzymes and separating the digested DNA on agarose or polyacrylamide gels.
2. CAPS analysis is versatile and can be combined with single strand conformational polymorphim (SSCP), sequence-characterized amplified region (SCAR), or random amplified polymorphic DNA (RAPD) analysis to increase the chance of finding a DNA polymorphism.
3. Michaels and Amasino (1998) proposed a variant of the CAPS method called CAPS based on SNPs.

Advantages

1. The genotyping throughput is moderately high.
2. It is a simple PCR assay. Markers are developed from the DNA sequences of previously mapped RFLP markers.
3. Most CAPS markers are co-dominant and locus specific.
4. No special equipment is needed to perform manual CAPS marker assays.
5. CAPS marker assays can be performed using semi-automated methods, *e.g.*, fluorescent assays on a DNA sequencer (*e.g.*, ABI377).
6. Start-up costs are low for manual assay methods.
7. CAPS assays can be performed using very small DNA samples (typically 50 to100 ng per individual).
8. Most CAPS genotypes are easily scored and interpreted.
9. CAPS markers are easily shared between laboratories.

Disadvantages

1. Typically, a battery of restriction enzymes must be tested to find polymorphisms.
2. Although CAPS markers still have great utility and should not be over looked, other methods have emerged as tools for screening locus-specific DNA fragments for polymorphisms, *e.g.*, SNP assays.
3. The development of easily scored and interpreted assays may be difficult for some genes, especially those belonging to multi-gene families.

Uses

This is straight forward way to develop PCR-based markers from the DNA sequences of previously mapped RFLP markers. Simple method that build on the investment of an RFLP map and eliminates the need for DNA blotting.

5. Simple Sequence Repeats (SSRs)

Simple sequence repeats (SSRs) or microsatellites are tandemly repeated mono-, di-, tri-, tetra-, penta-, and hexa-nucleotide motifs. SSR length polymorphisms are caused by differences in the number of repeats. SSR loci are individually amplified by PCR using pairs of oligonucleotide primers specific to unique DNA sequences flanking the SSR sequence. Jeffreys (1985) showed that some restriction fragment length polymorphisms are caused by VNTRs. The name "mini-satellite" was coined because of the similarity of VNTRs to larger satellite DNA repeats.

Advantages

1. SSR markers tend to be highly polymorphic.
2. The genotyping throughput is high.
3. This is a simple PCR assay.
4. Many SSR markers are multi-allelic and highly polymorphic.
5. SSR markers can be multiplexed, either functionally by pooling independent PCR products or by true multiplex-PCR.
6. Semi-automated SSR genotyping methods have been developed.
7. Most SSRs are co-dominant and locus specific.
8. No special equipment is needed for performing SSRs assays; however, special equipment is needed for some assay methods, *e.g.*, semi-automated fluorescent assays performed on a DNA Sequencer.
9. Start-up costs are low for manual assay methods (once the markers are developed).
10. SSR assays can be performed using very small DNA samples (~100 ng per individual).
11. SSR markers are easily shared between laboratories.

The difficulty of sharing markers (maintaining and distributing clones) has always been one of the key drawbacks of RFLP markers, particularly in species without a critical research mass.

Disadvantages

1. The development of SSRs is labor intensive.
2. SSR marker development costs are very high.
3. SSR markers are taxa specific.
4. Start-up costs are high for automated SSR assay methods.
5. Developing PCR multiplexes is difficult and expensive. Some markers may notmultiplex.

Uses

SSR markers are used for mapping of genes in eukaryotes.

6. Single Strand Conformational Polymorphisms (SSCPs)

SSCPs refer to DNA polymorphisms produced by differential folding of single-stranded DNA harboring mutations (Orita *et al.*, 1989). The conformation of the folded DNA molecule is produced by intra-molecular interactions and is thus a function of the DNA sequence. SSCP marker assays are performed using heat-denatured DNA on non-denaturing DNA sequencing gels. Special gels (*e.g.*, mutation detection enhancement gels) have been developed to enhance the discovery of single strand conformational polymorphisms caused by INDELs, SNPs, or SSRs.

Advantages

1. Simple PCR assay.
2. Many SSCP markers are multi-allelic and highly polymorphic.
3. Most SSCPs are co-dominant and locus specific.
4. No special equipment is needed.
5. Start-up costs are low.
6. SSCP marker assays can be performed using very small DNA samples (typically 10 to 50 ng per individual).
7. SSCP markers are easily shared between laboratories.
8. SSCP gels can be silver stained (no radioactivity).
9. The complexity of PCR products can be assessed and individual fragments can be isolated and sequenced.

Disadvantages

1. The development of SSCP markers is labor intensive.
2. SSCP marker development costs can be high.
3. SSCP marker analysis cannot be automated.

Uses

SSCPs have been widely used in human genetics to screen disease genes for DNA polymorphisms. Although SSCP analysis does not uncover every DNA sequence polymorphism, the methodology is straight forward and a significant number of polymorphisms can be discovered. SSCP analysis can be a powerful tool for assessing the complexity of PCR products (Slabaugh et al., 1997).

7. Heteroduplex Analysis (HA)

It refers to DNA polymorphisms produced by separating homo-duplex from hetero-duplex DNA using non-denaturing gel electrophoresis (White *et al.*, 1992) or partially denaturing high performance liquid chromatography (Wang *et al.*, 1999; Cho *et al.*, 1999). Single-base mismatches between genotypes produce hetero-duplexes; thus, the presence of hetero-duplexes signals the presence of DNA polymorphisms. Hetero-duplex analyses can be rapidly and efficiently performed on numerous genotypes before specific alleles are sequenced, thereby greatly reducing sequencing costs in SNP discovery and SNP marker development.

Advantages

1. Powerful method for SNP discovery.
2. Automated HA can be performed using HPLC.
3. Most hetero-duplex markers are co-dominant and locus specific.
4. HA can be performed using very small DNA samples (typically 10 to 50 ng per individual).
5. HA markers are easily shared between laboratories.

Disadvantages

1. Requires special equipment.
2. One protocol may not be sufficient for hetero-duplex analyses of different targets via HPLC.

Uses

Hetero-duplex analysis has been mostly used in human genetics to screen disease genes for DNA polymorphism. In plant breeding, it is used for detection of pathogens which are in latent stage and thus useful in selection of disease free plants. It is also useful in the discovery of single nucleotide polymorphism.

8. Single Nucleotide Polymorphism (SNP)

The variations which are found at a single nucleotide position are known as single nucleotide polymorphisms or SNP. Such variation results due to substitution, deletion or insertion. This type of polymorphisms have two alleles and also called bi-alleleic loci. This is the most common class of DNA polymorphism. It is found both in narural lines and after induced mutagenesis. Main features of SNP markers are given below.

1. SNP markers are highly polymorphic and mostly bialleleic.

2. The genotyping throughput is very high.
3. SNP markers are locus specific.
4. Such variation results due to substitution, deletion or insertion.
5. SNP markers are excellent long term investment.
6. SNP markers can be used to pinpoint functional polymorphism.
7. This technique requires small amount of DNA

Advantages

1. SNP markers are useful in gene mapping
2. SNPs help in detection of mutations at molecular level.
3. SNP markers are useful in positional cloning of a mutant locus.
4. SNP markers are useful in detection of disease causing genes.

Disadvantages

1. Most of the SNPs are bialleleic and less informative than SSRs.
2. Multiplexing is not possible for all loci.
3. Some SNP assay techniques are costly.
4. Development of SNP markers is labor oriented.
5. More (three times) SNPs are required in preparing genetic maps than SSR markers.

Uses

SNPs are useful in preparing genetic maps. They have been used in preparing human genetic maps. In plant breeding, SNPs have been used to lesser extent.

9. Expressed Sequence Tags [EST]

Expressed Sequence Tags (ESTs) are small pieces of DNA and their location and sequence on the chromosome are known. The variations which are found at a single nucleotide position are known. The term Expressed Sequence Tags (ESTs) was first used by Venter and his colleagues in 1991Main features of EST markers are given below.

1. ESTs are short DNA sequences (200-500 nucleotide long).
2. They are a type of sequence tagged sites (STS).
3. ESTs consist of exons only.

Advantages

1. It is a rapid and inexpensive technique of locating a gene.
2. ESTs are useful in discovering new genes related to genetic diseases.
3. They can be used for tissue specific gene expression.

Disadvantages

1. ESTs have lack of prime specificity.
2. It is a time consuming and labor oriented technique.
3. The precision is lesser than other techniques.
4. It is difficult to obtain large (> 6kb) transcripts.
5. Multiplexing is not possible for all loci.

Uses

ESTs are commonly used to map genes of known function. They are also used for phylogenetic studies and generating DNA arrays.

10. Sequence Tagged Sites [STS]

In genomics, a sequence tagged site (STS) is a short DNA sequence that has a single copy in a genome and whose location and base sequence are known. Main features of STS markers are given below.

1. STSs are short DNA sequences (200-500 nucleotide long).
2. STSs occur only once in the genome.
3. STS are detected by PCR in the presence of all other genomic sequences.
4. STSs are derived from c-DNAs.

Advantages

1. STSs are useful in physical mapping of genes.
2. This technique permits sharing of data across the laboratories.
3. It is a rapid and most specific technique than DNA hybridization techniques.
4. It has high degree of accuracy.
5. It can be automated.

TABLE 10-2: Characteristics of Commonly-Used Types of DNA Markers

Marker Type	*Lab Procedure*	*Reliability*	*Cost*
RFLP	Complex	High	High
RAPD	Simple	Low	Low
ISSR	Simple	Medium	Low
SSR	Simple	High	Low
AFLP	Complex	High	High
SCAR or STS	Simple	High	Low
CAPS	Moderate	High	Medium
InDel	Simple	High	Low

Disadvantages

1. Development of STS is a difficult task.
2. It is time consuming and labor oriented technique.
3. It requires high technical skill.

Uses

STS is the most powerful physical mapping technique. It can be used to identify any locus on the chromosome. STSs are used as standard markers to find out gene in any region of the genome. It is used for constructing detailed maps of large genomes.

Applications in Crop Improvement

DNA markers have several useful applications in crop improvement. The important applications are listed as follows:

1. DNA markers are useful in the assessment of genetic diversity in germplasm, cultivars and advanced breeding material.
2. DNA markers can be used for constructing genetic linkage maps.
3. DNA markers are useful in identification of new useful alleles in the germplasm and wild species of crop plants.
4. DNA markers are used in the marker assisted or marker aided selection. MAS has several advantages over straight selection.
5. DNA markers are useful in the study of crop evolution.

SUMMARY

Those characters which can be easily identifies are called marker characters. Markers related to variations in DNA fragments generated by restriction endonuclease enzymes are called DNA markers or genetic markers. Markers are of four types, *viz*.. (i) morphological, (ii) biochemical, (iii) cytological, and (iv) DNA markers. All these have been discussed. Properties of ideal DNA markers have been listed.

There are several types of DNA markers that are used in genetics and plant breeding for various purposes. The commonly used DNA markers include: Restriction Fragment Length Polymorphism (RFLP), Amplified Fragment Length Polymorphism (AFLP), Random Amplified Polymorphic DNA (RAPD), Cleaved Amplified Polymorphic Sequences (CAPS), Simple Sequence Repeat (SSR) Length Polymorphism, Single Strand Conformational Polymorphism (SSCP), Hetero-duplex Analysis (HA), Single Nucleotide Polymorphism (SNP), Expressed Sequence Tags [EST] and Sequence Tagged Sites [STS]. Advantages, disadvantages and uses of all these DNA markers have been discussed.

DNA markers have several useful applications in crop improvement. Important applications include: assessment of diversity, gene mapping, discovery of new genes, marker assisted selection and study of crop evolution.

QUESTIONS

1. **What is a marker? What are the demerits of morphological markers?**
2. **What are the advantages of molecular markers?**
3. **Describe important features of an ideal DNA marker.**
4. **Give a brief comparison of morphological markers and DNA markers.**
5. **Write short notes on the following:**

   (i) RFLP (ii) AFLP

   (iii) RAPD (iv) SSR
6. **Define the following terms:**

   (i) CAPS (ii) SSCP

   (iii) SNP (iv) EST

   (v) STS (vi) HA
7. **Give a comparison of Restriction Fragment Length Polymorphism (RFLP) and Amplified Fragment Length Polymorphism (AFLP).**
8. **What are the applications of DNA markers in crop improvement?**
9. **Define RFLP and AFLP markers and discuss their advantages, disadvantages and uses.**
10. **Define RAPD and SSR markers and discuss their merits, demerits and uses.**
11. **Define CAPS and SSCP markers and discuss their advantages, disadvantages and uses.**
12. **Define SNP and HA markers and discuss their advantages, disadvantages and uses.**
13. **Define EST and STS markers and discuss their merits demerits and uses.**

Chapter 11

Marker Assisted Selection

INTRODUCTION

Marker Assisted Selection is a process whereby a marker is used for indirect selection of a genetic trait of interest. It is also known as SMART breeding. [SMART=**S**election with **M**arkers and **A**dvanced **R**eproductive **T**echnologies]. MAS is a technique that does not replace traditional breeding, but can help to make it more efficient. It does not include the transfer of isolated gene sequences such as genetic engineering, but offers tools for targeted selection of the existing plant material for further breeding. MAS has already proven to be a valuable tool for plant breeders:

(i) It requires less investment, raises fewer safety concerns, respects species barriers, and is accepted by the public. MAS has high potential to meet challenges such as a changing climate, disease resistance or higher nutritional qualities.

(ii) Marker-assisted selection (MAS) is a modern plant breeding technique that can offer benefits to farmers developing climate or diseases resistant varieties, without the need for genetic engineering.

Main Features

Since smart breeding involves a far more precise knowledge, it is also known as precision breeding. Main features of SMART breeding are listed as follows:

1 **Principle:** Smart Breeding is a method based on the results of genomic research. Genomic research can supplant traditional time- and cost-intensive laboratory or field tests and can speed up the selection process.

2. **Use of Molecular Markers:** In smart breeding, when selecting the plants, the breeders do not rely on outward features (phenotype). Instead, plant breeders use molecular markers [molecular diagnostic tools - the so-called markers] to select parents for crossing or individual plants with desirable traits.

3. **Alternative to Transgenic Breeding:** In contrast to Genetic Engineering, Smart Breeding does not introduce any new genes into the genome. The resulting organisms are not transgenic. Smart breeding has been advanced as an alternative to transgenic plants as a way to produce plants that are resistant to various environmental problems. Smart Breeding which involves Marker-Assisted Selection, is a non-invasive biotechnology alterative to genetic engineering of plant varieties.
4. **No Effect of Environment:** Smart Breeding enables the detection of desired performance features irrespective of environmental influences.
5. **Rapid Method:** The analysis of the trait in question is a quick and easily automated process allowing a high throughput. Moreover, the existence of a trait can be analysed in earlier generations and in earlier development states (germ bud). Traditional methods can only decide upon resistance by carrying out infection trials in breeding nurseries or laboratories and by analysing yield and quality of the harvested crop which is very much time consuming.
6. **High Accuracy:** Smart breeding basically works in a similar way to traditional breeding. However, unlike traditional methods, in smart breeding the gene or gene variant responsible for a specific trait can be accurately identified using molecular biological procedures (DNA sequencing, PCR). It is then possible to test the offspring of a cross for the presence of the crossed gene, even before the actual trait is expressed by a changed external appearance. Only those plants which contain the desired gene are selected and grown further. Thus results of smart breeding are much more predictable.
7. **Purpose:** The purpose of Smart breeding is to introduce into crop plants genes from *e.g.* wild populations which confer characteristics of interest to breeders (disease resistance, fruit color or sugar content).

This technique was successfully used by Nachum Kedar in Israel in tomato breeding to produce a fruit that would ripen on the vine and remain firm in transit.

Marker Assisted Selection

Marker assisted selection (MAS) is a process whereby a marker (morphological, biochemical or one based on DNA/RNA) is used for indirect selection of a specific trait. Main points related to marker assisted selection are listed below.

(i) Marker assisted selection is also known as marker aided selection (MAS).

(ii) It is a process in which a marker (morphological, biochemical or one based on DNA/RNA variation) is used for indirect selection of a trait of interest.

(iii) A trait of interest may include productivity, disease resistance, abiotic stress tolerance, quality, *etc.*

(iv) This process is used both in plant and animal breeding.

(v) Marker assisted selection (MAS) is indirect selection process where a trait of interest is selected not based on the trait itself but on a marker linked to it.

(vi) MAS can be useful for traits that are difficult to measure, exhibit low heritability, and/or are expressed late in development.

MAS can be used as an aid to traditional phenotypic-pedigree-based selection systems.

Gene versus Marker

The gene of interest is directly related with production of protein(s) that produce certain phenotypes whereas markers should not influence the trait of interest but are genetically linked (remain together during segregation). In many traits genes are discovered and can be directly assayed for their presence with a high level of confidence. However, if a gene is not isolated marker's help is taken to tag a gene of interest. In such case there may be some false positive results due to recombination between marker of interest and gene (or QTL). A perfect marker would elicit no false positive results.

Selection for Major Gene Linked to Marker

The major genes control oligogenic or monogenic characters which are economically important. Such characteristics include disease resistance, male sterility, self-incompatibility, others related to shape, color, and surface of plants are often of mono- or oligogenic in nature. The marker loci which are tightly linked to major genes can be used for selection and are sometimes more efficient than direct selection for the target gene. Such advantages in efficiency may be due for example, to higher expression of the marker mRNA in such cases that the marker is actually a gene. Alternatively, in such cases that the target gene of interest differs between two alleles by a difficult-to-detect single nucleotide polymorphism, an external marker (be it another gene or a polymorphism that is easier to detect, such as a short tandem repeat) may present as the most realistic option.

Situation for Use of MAS

There are several situations where the use of molecular markers in the selection of a genetic trait would be rewarding. Such situations include (i) late expression of the trait, (ii) trait controlled by recessive gene, (iii) resistance to biotic stresses, and (iv) presence of epistasis. These are briefly discussed below.

1. **Late expression of a Trait**: MAS will be useful when the trait of interest is expressed late in plant development, like flower and fruit features.
2. **Control by Recessive Gene**: MAS will be useful when the character of interest is governed by recessive gene. It can identify both dominant and recessive genes.
3. **Resistance to Biotic Stress**: MAS will be useful in screening the material for resistance to diseases and insects especially in those areas field inoculation with the pathogen is not allowed for safety reasons. Moreover, it eliminates undesirable environmental effects.

4. **Epistasis**: MAS will be useful when the phenotype is affected by two or more unlinked genes (epistatis) for example, selection for multiple genes which provide resistance against diseases or insect pests for gene pyramiding.

The cost of genotyping (an example of a molecular marker assay) is reducing while the cost of phenotyping is increasing particularly in developed countries thus increasing the attractiveness of MAS as the development of the technology continues.

Steps Involved in MAS

The marker assisted selection consists of two important steps, *viz*.. (i) mapping of gene, and (ii) use of this information for marker assisted selection. These are discussed as follows:

(i) Gene Mapping

The first step is to map the gene or quantitative trait locus (QTL) of interest. This is done by using different techniques. Each gene mapping technique has some merits and demerits. Detailed discussion of all such techniques is beyond the scope of this discussion. Five types of populations, *viz*.. (i) recombinant inbred lilen(RILs), (ii) near isogenic lines (NILs), (iii) back cross, (iv) double haploids (DH), and (v) F_2 segregations are commonly used for gene mapping or gene tagging. Linkage between the phenotype and markers which have already been mapped is tested in these populations in order to determine the position of the QTL. Such techniques are based on linkage and are therefore referred to as linkage mapping. This consists of following steps:

1. Selection of parents with divergent alleles or contrasting traits.
2. Development of mapping populations.
3. Isolation of DNA.
4. Scoring of DNA markers such as RFLPs or AFLPs *etc.*

(ii) Marker Assisted Selection

The second important step is to use above information for marker assisted selection. Generally, the markers to be used should be close to gene of interest (<5 recombination unit or cM) in order to ensure that only minor fraction of the selected individuals will be recombinants. Generally, two markers are used in order to reduce the chances of an error due to homologous recombination. For example, if two flanking markers are used at same time with an interval between them of approximately 20cM, there is higher probability (99 per cent) for recovery of the target gene. This consists of following steps:

1. Indirect selection using molecular markers, and
2. Correlation of DNA markers with morphological markers.

Various type of DNA markers are used in marker assisted selection. The commonly used markers include RFLPs, RAPDs, AFLP, SNPs, SCARs, microsatellites *etc.*

Single Step MAS and QTL Mapping

In contrast to two-step QTL mapping and MAS, a single-step method for breeding typical plant populations has been developed In such an approach, in the first few breeding cycles, markers linked to the trait of interest are identified by QTL mapping and later the same information is used in the same population. In this approach, pedigree structure are created from families that are created by crossing number of parents (in three-way or four way crosses). Both phenotyping and genotyping is done using molecular markers mapped the possible location of QTL of interest. This will identify markers and their favorable alleles. Once these favorable marker alleles are identified, the frequency of such alleles will be increased and response to marker assisted selection is estimated. Marker allele(s) with desirable effect will be further used in next selection cycle or other experiments.

High-throughput Genotyping Techniques

Recently high-throughput genotyping techniques are developed which allows marker aided screening of many genotyes. This will help breeders in shifting traditional breeding to marker aided selection. One example of such automation is using DNA isolation robots, capillary electrophoresis and pipetting robots. One of recent example of capillary system is Applied Biosystems 3130 Genetic Analyzer. This is the latest generation of 4-capillary electrophoresis instruments for the low to medium throughput laboratories.

Applications of Markers

In most major crops, various linked markers exist to screen key traits. Numerous markers have been mapped to different chromosomes in several crops including rice, wheat, maize, soybean and several others. These markers have been used in diversity analysis, parentage detection, DNA fingerprinting, and prediction of hybrid performance. Molecular markers are useful in indirect selection processes, enabling manual selection of individuals for further propagation. In addition to these, applications of MAS in backcrossing and gene pyramiding are discussed below.

1 Marker Assisted Backcross

A minimum of five or six backcrosses are required to transfer a gene of interest from a donor (may not be adapted) to a recipient (recurrent adapted) cultivar. The recovery of the recurrent genotype can be accelerated with the use of molecular markers. If the F_1 is heterozygous for the marker locus, individuals with the recurrent parent allele(s) at the marker locus in first or subsequent backcross generations will also carry a chromosome tagged by the marker.

2. Marker Assisted Gene Pyramiding

Gene pyramiding has been proposed and applied to enhance resistance to disease and insects by selecting for two or more than two genes at a time. For example in rice such pyramids have been developed against bacterial blight and blast. The advantage of use of markers in this case allows to select for QTL-allele-linked markers that have same phenotypic effect.

Advantages of MAS

Marker-assisted selection is the most widely used application of DNA markers. Once traits have been mapped and a closely linked marker has been found, it is possible to screen large numbers of samples for rapid identification of progeny that carry desirable characteristics. MAS has several advantages such as (i) high speed, (ii) consistency of results, (iii) bio-safety, (iv) high efficiency, and (v) QTL mapping. These are briefly discussed below.

1. **Speed:** It is a rapid method of selection of desirable plants. DNA can be extracted from tissue from the first leaves or the cotyledons of a plant. Trait information can be discovered with markers prior to pollination allowing more informed crosses to be made.
2. **Consistency:** The results of MAS are consistent, because use of markers eliminates the impact of environmental variation that often complicates phenotypic evaluation.
3. **Biosafety:** MAS is a safe method of screening. When we use MAS, there is no need of introducing into breeding populations. Particularly for livestock breeding this delivers a very important level of bio-safety.
4. **Efficiency:** MAS has very high efficiency. Screening of progeny in early stages allows a breeder to reject undesirable genotypes from the program more quickly. Most breeding programs that use markers still evaluate the same number of plants in the field however the level of genetic quality is vastly increased because of the early-stage screening that has been carried.
5. **Complex traits:** DNA markers permit mapping of polygenic traits or quantitative trait loci [QTL] which is not possible through conventional plant breeding techniques.

Disadvantages of MAS

Marker assisted selection has some disadvantages such as (i) expensive technique, (ii) requires technical skill, (iii) a laborious work, and (iv) may sometimes lead to health hazards. These are briefly discussed below.

1. **Expensive Technique**: MAS is an expensive technique, because this technique requires very costly equipments, glassware and chemicals; sophisticated laboratory.
2. **Technical skill:** This technique requires well trained man power for handling of equipments, isolation of DNA molecules and study of DNA markers.
3. **Laborious Work**: The detection of various DNA markers [RFLP, AFLP, RAPD, SSR, SNP *etc.*] is a laborious and time consuming work. For such markers huge plant breeding population has to be screened to get meaningful results.

4. **Health Hazards**: Some of the DNA marker techniques involve use of radioactive isotopes in labeling of DNA, which may lead to serious health hazards. Now non radioactive labeling agents are also available.

SUMMARY

Marker Assisted Selection [MAS] is a process whereby a marker is used for indirect selection of a genetic trait of interest. It is also known as SMART breeding.

MAS is a technique that does not replace traditional breeding, but can help to make it more efficient. It does not include the transfer of isolated gene sequences such as genetic engineering, but offers tools for targeted selection of the existing plant material for further breeding. MAS has been a valuable tool for plant breeders:

Marker assisted selection refers to selection for specific alleles (which affect a trait of interest) using genetic marker. In other words, Marker assisted selection (MAS) is a process whereby a marker (morphological, biochemical or one based on DNA/RNA) is used for indirect selection of a specific trait. Markers are of four main types, *viz*.. (i) morphological, (ii) biochemical, (iii) cytological, and (iv) DNA markers.

Morphological markers are related to variation in shape, size, colour and surface of various plant parts. Biochemical markers are related to variation in proteins and amino acid banding pattern.

Cytological markers that are related to variation in chromosome number, shape, size and banding pattern are referred to as cytological markers. In other words, such markers refer to the chromosomal banding produced by different stains; for example, G banding.

DNA marker refers to any unique DNA sequence which can be used in DNA hybridization, PCR or restriction mapping experiments to identify that sequence. It can be identified by a range of molecular techniques such as RFLPs, RAPDs, AFLP, SNPs, SCARs, microsatellites *etc*. DNA markers are also known as molecular markers or genetic markers.

Comparison of morphological markers and DNA markers has been presented. Characteristics of an ideal DNA marker are given and steps involved in marker assisted selection have been discussed. Applications, advantages and disadvantages of Marker Assisted Selection have been described.

QUESTIONS

1. **Define MAS and explain its important features.**
2. **Explain the situations when marker assisted selection would be rewarding.**
3. **Describe briefly disadvantages of marker assisted selection.**
4. **What are various steps involved in marker assisted selection?**

5. **Define the following terms:**
   (i) Morphological markers
   (ii) Biochemical markers
   (iii) Cytological markers
   (iv) DNA markers
6. **Describe briefly marker assisted selection in relation to backcrossing and gene pyramiding**
7. **Discuss briefly various applications of marker assisted selection in crop improvement.**

Chapter 12

Transgenic Breeding

INTRODUCTION

The technique or process of developing transgenic individuals is referred to as transgenic technology. It is also known as gene technology, transformation technology, genetic engineering and recombinant DNA technology.

Transgenic Plants

The genetically engineered plants are called transgenic plants. In other words, plants which are developed by the technique of biotechnology are referred to as transgenic plants. Main features of transgenic plants are listed as under.

1. Transgenic plants contain transgene or foreign gene of special significance.
2. Transgenic plants are developed through plant biotechnology.
3. In the development of transgenic plants, the sexual process is not involved.
4. The transgenic plants are recovered at a very low frequency.
5. Transgenic plants are developed to tackle those problems which cannot be solved by conventional plant breeding techniques.

Advantages of Transgenic Technology

Transgenic technology has several advantages such as rapid method, free gene transfer, single gene transfer, direct gene transfer and solution to difficult problems. These are briefly discussed as follows:

1. **Rapid Method:** It is an effective and rapid method of crop improvement. It takes 3-4 years for developing new cultivars/hybrids against 10-15 years taken by conventional methods.
2. **Free Gene Transfer:** It permits gene transfer across the species, genera, family even from unrelated organisms.

3. **Single Gene Transfer:** Gene technology permits transfer of one or two genes from donor parent to the recipient parent. In conventional hybridization method, hundreds of genes are transferred to the recipient parent. Many of transferred genes are undesirable. Elimination of such genes requires repeated backcrossing to the recipient parent which takes 4-5 years.
4. **Direct Gene Transfer:** Gene technology permits direct gene transfer into the recipient parent bypassing sexual process. In other words, there is no need of union of male and female gametes in gene technology. The gene of interest can be directly inserted into the cell of recipient parent.
5. **Solution to Difficult Problems**: It helps in providing solution to those problems that cannot be solved by conventional methods of breeding such as bollworm resistance in cotton.

Steps Involved

The transgenic plants are developed by the technique of genetic engineering. Seven important steps, *viz*.. gene identification, gene transfer, identification of transferred gene, regeneration, gene expression, proper integration and field trials, are involved in the development of transgenic plants. These steps are discussed below using example of Bt. Cotton.

(i) **Gene identification**: First the gene of interest is identified. In case of cotton, the bollworm resistant gene is found in the soil bacterium *Bacillus thuringiensis*. The toxin producing gene is isolated from this bacterium and then its multiple copies are made[cloning].

(ii) **Gene transfer**: The second important step is to transfer the bollworm resistant gene in to cotton plant. There are four important methods of foreign gene (DNA) transfer in crop plants *viz*.. Plasmid method, particle bombardment, direct DNA uptake and micro-injection. These methods are also known as systems of DNA delivery for genetic transformation. The soil borne bacterium *Agrobacterium tumefaciens* is used for development of transgenic plants. Currently, two DNA delivery system *viz*.. (i) *Agrobacterium* mediated gene transfer, and (ii) bombardment of cells with plasmid DNA coated particles, are widely used for development of transgenic (genetically engineered) plants in cotton. More than 37 transgenic plants have been developed in cotton so far by these two methods.

(iii) **Identification of Transformed cells**: After gene delivery, the transformed cells are identified. The presence of transgene or gene of interest is detected by several methods. The commonly used techniques of detecting transgenes are (a) selectable marker gene, (b) Polymerase chain reaction [PCR], (c) Enzyme Linked Immuno-sorabant Assay [ELISA], (d) Southern blot technique, (e) Northern blot technique, and (f) Western blot technique. Of these, the selectable marker gene is is simple and widely used.

IDENTIFICATION OF GENE OF INTEREST

↓

GENE ISOLATION

↓

GENE CLONING

↓

TRANSFER TO HOST PLANT

↓

IDENTIFICATION OF TRANSGENIC PLANTS

↓

HARDENING OF REGENERATED PLANTS

↓

SMALL SCALE FIELD TRIALS

↓

LARGE SCALE FIELD TRIALS

FIGURE 12-1: Steps Involved in Development of Transgenic Plants.

A selectable marker is attached to the target gene for easy identification [of the gene of interest]. Generally an antibiotic resistant gene is used as selectable marker to confirm whether new gene has been inserted in the host cells or not. The Kanamycin resistant gene is used for identification of transformed cells. The lethal concentration of Kanamycin is put in the plasmid mixed suspension of protoplast culture. The non-transformed cells will die and only transformed protoplasts will survive. Thus transformed cells are identified.

(iv) **Regeneration**: The next important step is regeneration ability from protoplasts, callus or tissues. The transgenic cells, callus or protoplasts are regenerated in to whole plant with the help of tissue culture.

(v) **Gene expression**: The trans-gene inserted in to host plant should have expression of the desired level,

(vi) Proper integration: The gene should be integrated properly so that it is carried from generation to generation by usual means of reproduction.

(vii) Field Trials: First the transgenic plants are tested in the laboratory for bio-safety such as allergenicity and toxicity with animals like rats, rabbits, poultry, goats, *etc.* Transgenic plants which passed by the regulatory authority are evaluated in multi-location trials for three years to test the performance of the gene of interest. Superior genotypes with stable performance are released for cultivation.

Transgenic vs Conventional Breeding

The genetic improvement of crop plants through the application of agricultural biotechnology [transgenic plants] is called transgenic plant breeding. Transgenic crop breeding differs in many ways from conventional methods of crop breeding. Comparison of transgenic technology and traditional plant breeding is presented in the Table 12.1.

TABLE 12-1: Comparison of Gene Technology and Traditional Breeding Methods

Sl.No.	*Particulars*	*Gene Technology*	*Traditional Breeding*
1	Used in	Single celled organisms	Used in multi-cellular organisms
2	Sexual process	Not Involved	Involved
3	Occurrence in nature	Seldom	Common
4	Crossing barriers	Not present	Present
5	Technique used	Genetic engineering	Conventional breeding methods
6	Methods used	Agrobacterium based method, gene gun, micro injection and direct DNA uptake, *etc.*	Pedigree, bulk breeding, single seed descent, back cross breeding, *etc.*
7	Applications	In all crops	In all crops
8	Bio-safety measures	Required	Not required
9	End product	Transgenic plants	Non transgenic plants
10	Bioethical measures	Required	Not required
11	Evolution of super weeds	Possible	Not possible
12	Organic farming of end product	Not permitted	Permitted
13	Accuracy of method	Very high	Moderate to high
14	Time required for release of new variety	4-5 years	10-12 years
15	Technical skill required	Very high	Moderate
16	Laboratory required	Sophisticated	Ordinary
17	Initial cost of infra structure	Very high	Low
18	Direct single gene transfer	Possible	Not possible
19	Transgene	Involved	Not involved

Applications

Transgenic technology has several practical applications. Important applications of transgenic technology include improvement in yield, quality of food products and resistance to biotic and abiotic stresses. All these aspects are briefly discussed as follows:

1. **Improvement in Yield:** Transgenic technology plays important role in increasing the productivity of food, fibre and vegetable crops ensuring food security, which is essential for international peace and stability. Thus, it is an important means to fight hunger. The transgenes are not generally yield enhancing genes. The increase in yield or productivity is achieved by controlling losses caused by various insects, diseases and abiotic factors. Gene technology is expected to keep pace in food production with increasing world population.
2. **Resistance to Biotic Stresses:** Biotic stress refers to adverse effects on crop growth and yield by biotic factors such as insects, diseases and parasitic weeds. In crop plants, heavy yield losses are caused every year due to insect and disease attack. Moreover, insecticides and pesticides which are used to control insects and diseases are expensive and have adverse effects on other beneficial organisms (parasites and predators). Gene technology has played key role in developing insect resistant cultivars in several crops such as bollworm resistant cultivars in cotton and stem borer resistant cultivars in maize. Moreover, the technology is eco-friendly.
3. **Herbicide Resistance:** In crop plants, weeds cause heavy yield losses and also adversely affect quality of the produce. The genetic resistance is the cheapest and the best way of solving this problem. Gene technology has been used to develop herbicide resistant cultivars in cotton, maize, wheat, tobacco, potato, tomato, rapeseed, soybean, flax, *etc.* In these crops, cultivars resistant to glyphosate, gluphosinate and some other herbicides have been developed.
4. **Resistance to Abiotic Stresses**: Abiotic stress refers to adverse effects on crop growth and yield by abiotic factors such as drought, soil salinity, soil acidity, cold, frost, *etc.* The cold resistant genotypes have been developed in tobacco. Efforts are being made to develop drought and salinity resistant cultivars in many crops.
5. **Improvement in Quality:** The quality is adjudged in three ways, *viz..,* nutritional quality, market quality [keeping quality] and industrial quality. Gene technology has helped in improving these qualities in different crops. For example, the ripening and softening in tomato has been delayed. It is desirable for safe transport and storage. This has been achieved by manipulating the genes that encode the enzyme responsible for ripening [ethylene forming enzymes and softening [polygalactonase]. Other quality improvement include non-browning potato, starch composition of wheat flour, carotene content in rice and improved oil content in oilseed crops.

6. **Industrial Products:** Gene technology has great potential for the production of biodegradable plastics, obtaining therapeutic proteins, pharmaceuticals and edible vaccines from transgenic plants. It may also help in producing biodiesel or petroleum products.
7. **Accurate and Rapid Technique**: Gene technology is rapid and highly accurate method of crop improvement. The development of cultivars by this technique takes 4-5 years against 10-12 years taken by conventional [hybridization] method. Moreover, this is a highly reliable technology.
8. **Free Gene Transfer:** Gene technology permits free gene transfer across species and genera whether related or unrelated. The gene of interest can be transferred from micro-organism to higher plants and even from animals to plants. Thus, it overcomes natural barriers of gene transfer.
9. **Bioremediation:** It is possible to develop plants which can be used for bio-remediation of sick soils.

Practical Achievements

Plant biotechnology has played significant role in improving yield, in some cases food quality and reducing use of pesticides or particularly harmful herbicides. Practical achievements of transgenic technology include improvement in biotic and abiotic resistance, quality and induction of male sterility. Some examples of such achievements are cited below.

1. **Improvement in Resistance:** It includes insect resistance, disease resistance, salinity resistance and herbicide resistance. Such resistance has been achieved in several crops as mentioned below.
   - (i) **Insect Resistance**: It includes bollworm resistance in cotton, European borer resistance in maize, Resistance to these insects has been transferred from *Bacillus thuringiensis*.
   - (ii) **Disease Resistance**: Resistance to Pierce disease in grapes has been developed at the University of Florida.
   - (iii) **Salinity Resistance**: Salinity resistant tomato has been developed at the University of California and at the University of Toronto.
   - (iv) **Herbicide Resistance**: It has been achieved in cotton, Flax, Soybean, Wheat, Potato, Tomato, Rapeseed, *etc.* The gene for herbicide resistance has been transferred from Steptomyces to wheat, potato and tomato; and from microbial gene in cotton, soybean and linseed.
2. **Improvement in Quality**: Quality improvement has been achieved in several crops. Some examples are cited below.
   - (i) **High Starch Potato**: In potato starch contents have been increased by transferring gene from human intestine bacteria [*Escherichia coli*].
   - (ii) **High Protein Potato and Alfalfa**: Protein contents have been improved in potato and alfalfa. In potato, the high protein gene has been transferred from Amaranthus and serum albumin gene from human. In alfalfa ova-albumin gene has been transferred from chicken.

(iii) **High Methionine Soybean**: In soybean, methionine content has been enhanced by transferring gene from a bacteria [*Bertholletia excels*].

(iv) **High Carotene Rice**: In rice, high carotene content gene has been transferred from Daffodils and **Golden rice** has been developed. The golden rice is rich in vitamin A content.

(v) **Anti-cracking Tomato**: In tomato, anti-cracking gene has been transferred from winter flounder fish. This has increased the transportation and keeping quality of tomato.

(vi) **Freezing Resistant Tobacco**: In tobacco, freezing resistant gene has been transferred from winter flounder fish and cold resistant gene from *Arabidopsis thaliana*.

3. **Male Sterility:** The male sterility has been developed in rapeseed through the application of agricultural biotechnology. The *barnase* and *barstar* system has been developed in this crop. The male sterility gene has been transferred from *Bacillus amyloliquefaciens*.

Probable Risks

There are some probable risks of plant biotechnology such as adverse effects, herbicide resistance, ethical issues and antibiotic resistance. These are explained as follows:

1. **Adverse Effects:** Consumption of transgenic food and vegetables may have adverse effects on human and animal health such as toxicity, allergic effects. Six transgenic crops, *viz.*. soybean, maize, cotton, potato and sugar beet are grown in Australia for human consumption. However, no adverse effects have been reported so far. The effects of GM crops on health and environment are uncertain. The cultivation of such crops can have unintended adverse effects on both animal health and the environment.
2. **Herbicide Resistance:** There is fear that natural crossing of herbicide resistant cultivars with wild relatives may lead to development of herbicide resistant weeds. However, no such confirmed case has been reported so far.
3. **Ethical Issue:** The transfer of gene from animal such as from fish to plants will have ethical problems.
4. **Antibiotic Resistance**: Wide spread cultivation of transgenic plants may lead to development of antibiotic resistance by insect pests and other pathogens.

Future Scope

It is expected that plant biotechnology [GM crops] can revolutionize world agriculture, particularly in developing countries. This technology will substantially improve food security, reduce malnutrition, and increase rural income, and in some cases even reduce environmental pollutants. If the GM crops are free of adverse effects on health and environment, they have the potential to provide benefits to farmers and consumers around the globe. Such crops will reduce the use of potentially harmful chemicals or scarce water supplies for agriculture. It can then indeed become a true Gene Revolution. In future, several wonderful achievements are likely to be made through the application of plant biotechnology in product quality and other areas. Some examples are cited below.

1. Development of protein packed potatoes.
2. Development of Nicotine free tobacco.
3. Development of vitamin E rich Canola.
4. Development of banana with inactivated viruses which cause cholera, hepatitis B and diarrhea.
5. Improvement in the storage and keeping quality of fruits and vegetables.
6. Production of oral vaccines and other pharmaceuticals in plants.
7. Production of biodegradable polymers for the plastic industry.
8. Plants that can be used for bioremediation of polluted soils.
9. Forest trees with enhanced growth and improved timber properties.
10. Production of enhanced quality of chemicals for chemical industry.

In future, to achieve agricultural revolution through gene technology, the following points should be given due importance.

1. **Affordable Technology:** The new technology should be affordable by the farmers of developing-world. Moreover, the farmers must understand how to use them.
2. **Research by Public Sector:** There is a need for larger investments in research in the public sector. Partnerships between the public and private sectors can result in more efficient production of GM crops that are useful to the developing world and can expand the accessibility of those crops and their associated technologies to developing-world farmers.
3. **Due Importance to Agriculture:** Agricultural development must be given due importance from a policy perspective in both donor and recipient nations. In view of increasing global population, agricultural development is more necessary than ever to eliminate malnutrition and prevent famine, particularly in sub-Saharan Africa. GM crops seem to be effective means for addressing these problems. The policymakers should work jointly on such issues.
4. **Proper Regulatory System:** Policymakers in the developing world must set regulatory standards that take into consideration the risks as well as the benefits of foods derived from GM crops. This goal is crucial to the cooperation of the many stakeholders that are affected by GM crops and also for the sustainability of the GM crop movement in the near future. Without regulations that explicitly take into account potential benefits to both farmers and consumers, those nations that might stand to benefit most

from GM crops may be discouraged from allowing them to be planted. Revised regulations on genetically modified crops must accompany widespread collective policy efforts to revitalize agricultural development.

SUMMARY

The technique or process of developing transgenic individuals is referred to as transgenic technology. It is also known as gene technology, transformation technology, genetic engineering and recombinant DNA technology.

Transgenic technology has several advantages such as rapid method, free gene transfer, single gene transfer, direct gene transfer and solution to difficult problems. These are briefly discussed.

The genetically engineered plants are called transgenic plants. In other words, plants which are developed by the technique of biotechnology are referred to as transgenic plants. Main features of transgenic plants are listed.

The genetic improvement of crop plants through the application of agricultural biotechnology [transgenic plants] is called transgenic plant breeding. Transgenic crop breeding differs in many ways from conventional methods of crop breeding. Comparison of transgenic technology and traditional plant breeding is presented in the tabular form.

Transgenic technology has several practical applications. Important applications of transgenic technology include improvement in yield, quality of food products and resistance to biotic and abiotic stresses. All these aspects have been discussed.

Plant biotechnology has played significant role in improving yield, in some cases food quality and reducing use of pesticides or particularly harmful herbicides. Practical achievements of transgenic technology include improvement in biotic and abiotic resistance, quality and induction of male sterility. Some examples of such achievements have been cited. There are some probable risks of plant biotechnology such as adverse effects, herbicide resistance, ethical issues and antibiotic resistance. These are have been explained. The future scope of plant biotechnology is presented.

QUESTIONS

1. **What is gene biotechnology? Describe advantages of transgenic technology.**
2. **What are limitations of transgenic technology?**
3. **Explain briefly the comparison of transgenic technology and conventional plant breeding.**
4. **What are various applications of transgenic technology with examples?**
5. **Give a brief account of the practical achievements of transgenic technology.**
6. **Explain briefly the probable risks of transgenic technology.**
7. **Describe the role of transgenic technology in quality improvement.**

8. Explain briefly the future scope of transgenic technology.

9. Write short notes on the following:

(i) Golden Rice

(ii) Anti-cracking tomato

(iv) Freezing resistant tobacco

(iv) Transgenic male sterility

Chapter 13

Transgenic Male Sterility

INTRODUCTION

Transgenic male sterility was first reported by Mariani *et al.*, in 1990 in rapeseed. A condition in which either pollen is absent or non-functional in flowering plants is referred to as male sterility. The main points related to male sterility are listed as follows:

1. Male sterility is an important outbreeding device which prevents autogamy and permits allogamy; In other words, male sterile plants produce seed only on cross pollination with functional pollen of other plants.
2. Presence of male sterility leads to heterozygosity in a species as it promotes outbreeding and reduces homozygosity due to elimination of inbreeding.
3. Male sterility results from the action of nuclear genes or cytoplasmic genes or both.
4. Male sterility is caused due to pollen or anther abortion.
5. Male sterility occurs in nature through spontaneous mutations as well as can be induced artificially by chemical or physical mutagens. Ethidium bromide has been found effective in inducing cytoplasmic male sterility in some plants.
6. Male sterility can be observed in all diploid species of crop plants both wild and cultivated if properly investigated.
7. In crop plants, three types of male sterility, *viz*.. (i) genetic (ii) cytoplasmic, and (iii) cytoplasmic genetic are found.

Transgenic Male Sterility

In addition to above three types of male sterility, there is another class of male sterility what we call transgenic male sterility. The foreign gene or transgene is also used for induction of male sterility. The male sterility which is induced by the technique of genetic engineering is referred to as transgenic male sterility. The gene *barnase* is the first transgene that was used to produce male sterility by Mariani *et al.*, in 1990. It has an effective fertility restoration system in *barstar*. Transgenic male sterility comes under genetic male sterility and is heritable. This type of sterility has been induced in crop plants like tobacco, rapeseed, maize, *etc.* The main features of transgenic male sterility related to source of gene, types of genes involved, gene integration, nature of gene, material included, maintenance and uses are briefly discussed as follows:

1. **Source of Gene:** The gene for inducing transgenic male sterility is generally used from micro-organisms. Transgenic male sterility has been induced in tobacco and rapeseed by transferring gene from *Bacillus amyloliquefaciens*.
2. **Types of genes Involved:** Transgenic male sterility is controlled by foreign nuclear genes. In this system two types of transgenes, *viz..*, *barnase* and *barstar* are used. The gene *barnase* causes male sterility and the gene *barstar* restore the fertility. Both these genes have been utilized from the bacterium *Bacillus amyloliquefaciens*. The male fertility of *barnase* male sterile is restored by another gene, *barstar*, of the bacterium *B. amyloliquefaciens*.
3. **Gene Integration:** The gene *barnase* is integrated in male sterile line called A line and the gene *barstar* is integrated in restorer line, called R line. Thus, it is a two line system. The cross between A and R line yields fertile F1.
4. **Nature of Gene:** The transgenic male sterility is controlled by a dominant gene called *barnase*. The male sterile plants are always heterozygous. In other words, the transgenic male sterility reported so far is controlled by one dominant gene (*barnase*). The gene *barstar* is responsible for restoration of fertility in the hybrid developed using *barnase* male sterile line. For commercial exploitation, a dominant genetic male sterility system must have the following important features:
   (i) Efficient fertility restoration system,
   (ii) Easy maintenance of male sterile lines,
   (iii) Easy elimination of male fertile plants from the male sterile line,
   (iv) Lack of adverse effects on other traits,
   (v) Stable male sterile phenotype over different environments and genetic backgrounds, and
   (vi) Satisfactory performance of the F_1 hybrids.

   The *barnase*-bar/*barstar transgenic male sterility system* appears to satisfy above mentioned requirements.
5. **Material Included**: This system consists of three components, *viz..*, *A* line, B line and *R* line. Description of these lines is presented as follows:
   (i) A line: It refers to genetic male sterile line (*barnase*) which is used as female parent in hybrid seed production. It is a heterozygous line (*barnase*/-). A male-sterile (ms) line is a group of plants of a plant

variety, which are all male-sterile due to the presence of a particular male-sterility gene say *barnase* in this case. A male sterile plant contains a foreign "male-sterility locus" which contains the male-sterility gene S which when expressed in cells of the plant makes the plant male-sterile. Such a male-sterility gene is always a dominant allele at such a foreign male-sterility locus.

(ii) **B line:** It is a maintainer line. It is crossed with A line to maintain the male sterility.

(iii) **R line (barstar):** This line is used to store fertility in the F_1 hybrid. A restorer line is a group of plants of a plant species which all contain the a fertility restorer gene say *barstar* in this case at the same genetic locus. Male fertility is restored to a male-sterile line by introducing the fertility restorer gene in the ms line, *e.g.* by crossing of the plants of the ms line with plants of the restorer. The cross of the fertility restorer line with male-sterile line restores the male fertility in the F_1 progeny. Such plants are obtained from a cross between a male-sterile plant and the restorer plant.

5. **Maintenance**: The three components of transgenic male sterility are maintained separately as follows:

(i) **A line:** The transgenic male sterile line (A) is heterozygous (*barnase*/-) which is maintained by crossing male sterile plants (*barnase*/-) with normal male fertile lines (-/-) in each generation. From such cross only 50 per cent of the progeny will be male sterile (*barnase*/-) and the rest 50 per cent will be male fertile. In a hybrid seed production program, the male fertile plants present in the male sterile line are identified and eliminated.

TABLE 13-1: Comparison of Conventional Genetic Male Sterility and Transgenic Male Sterility

Sl.No.	*Particulars*	*Conventional GMS*	*Transgenic Male Sterility*
1	Gene involved	Usually recessive	Usually dominant
2	Source of male sterility gene	Plants	Micro-organisms
3	Method of development	Conventional	Genetic engineering
4	Component lines	Two A and B	Three A, B and R
5	Cost involved	Low	High
6	Technical skill required	Medium	High
7	Effect of environment	Sometimes observed	Absent
8	Applications	Wide	Limited
9	Identification	Possible after anthesis	Possible at seedling stage

(ii) **R Line:** Transgenic plants expressing *barstar* are male fertile without any phenotypic effect, and are easily maintained in the homozygous state by self-pollination. When a homozygous *barstar* male fertile line is crossed with a *barnase* male sterile line, all the progeny plants are male fertile since *barstar* gene product effectively inhibits the *barnase* RNase in *barnase-bar/barstar* plants.

(iii) B Line: This is a normal homozygous male fertile line which is maintained by self- pollination.

6. **Uses**: Transgenic male sterility can be used for the production of hybrid seed in both seed propagated and vegetatively propagated species.

Identification of 'A' Line

The male sterile line is identified with the help of another gene called *bar* gene. The bar gene has been isolated from *Streptomyces.* The *barnase* gene is linked with the *bar* gene of *Streptomyces.* The *bar* gene confers resistance to the herbicide phosphinothricin. When such male sterile (*barnase-bar/-*) plants are maintained by crossing with normal male fertile plants (-/-), all the male sterile progeny (*barnase-bar/-*) are resistant to the herbicide, while all the male fertile plants (-/-) are herbicide susceptible. The male fertile plants are, therefore, easily eliminated by the herbicide spray at an early stage of plant growth.

Transfer of Sterility and Restorer Genes

Initially, the male sterility foreign gene (*barnase*) and male sterility gene barstar are transferred to A line and R line by the technique of genetic engineering and the transformed cells are regenerated into a mature plant. Subsequently, the resulting transformed plants can be used in a conventional breeding scheme to produce more transformed plants with the same characteristics or to introduce the male-sterility gene, or the fertility-restorer gene in other varieties of the same or related plant species. Thus the male-sterility gene, or the fertility-restorer gene when introduced into a particular line or variety of a plant species can always be introduced into any other line or variety by backcrossing. The male sterility and fertility restorer genes are transferred to female and male parent by separate backcross program.

Barnase is the ribonuclease which is secreted by *Bacillus amyloliquefaciens* and *barstar* is the inhibitor of *barnase* that is produced by the same microorganism. The *barnase* gene can be used for the construction of a sterility gene while the *barstar* gene can be used for the construction of a fertility-restorer gene. Experiments in different plant species, *e.g.* oilseed rape, have shown that a chimeric *barstar* gene can fully restore the male fertility of male sterile lines in which the male sterility was due to the presence of a chimeric *barnase* gene.

The "restorer capacity refers to the ability and efficiency of a restorer line to restore male fertility efficiently to a wide variety of ms lines. It is directly related to the amount of barstar produced in the stamen, particularly in the tapetum cells, of the restorer plants. Restorer capacity is important because an efficient restorer line can be used for the restoration of male fertility to a wide variety of male-sterile lines.

Plant species in which the synthetic barstar DNAs are used include especially oilseed rape, maize and rice. Restorer genes comprising the synthetic barstar DNAs may be used for producing restorer lines in different plant species but are particularly useful for producing restorer lines in cereals, especially in corn, rice and wheat.

Advantages

1. Early identification of male sterile plants is very easy.
2. Transgenic male sterility can be used for the production of hybrid seed in both seed propagated crops and vegetatively propagated species.
3. Transgenic male sterility is stable and provides a complete and usable male sterility-restorer system.
4. Transgenic male sterility generally does not have undesirable agronomic characters.

Disadvantages

There are four main disadvantages of transgenic male sterility.

1. Transgenic male sterility is induced by the technique of genetic engineering which is an expensive method.
2. This type of male sterility has been used in very few crops so far.
3. Crosses between male sterile plants (*barnase*) and normal male fertile plants are made every year for maintaining the transgenic male sterility.
4. Transgenic male sterility requires more area and labor, because the breeder has to maintain three types of populations, *viz..*, *A*, *B* and *R* lines.

Applications

For hybrid seed production, it is necessary to ensure that fertility of the hybrid plants is fully restored. In systems in which the male sterility is under genetic control this requires the existence and use of genes that can restore male fertility. The development of hybrid cultivars is mainly dependent on the availability of suitable and effective sterility and restorer genes. For the production of hybrid seed the male-sterile line is also called the female or first parent line, and the male-fertile (restorer) line is also called the male or second parent line.

Transgenic male sterility/fertility system has shown commercial promise in maize and oilseed rape. Now it is also available in cauliflower, chicory, tomato and wheat. In India, transgenic lines expressing *barnase* and *barstar* genes have been developed in *Brassica juncea.* These lines are stable and provide a complete and usable male sterility-restorer system. This *barnase-barstar* system is expected to promote hybrid development in *Brassica juncea.*

Antisense Gene Approach

Antisense refers to short DNA or RNA sequences, termed oligonucleotides, which are complementary to a specific gene sequence. The goal is to alter specific gene expression resulting from the binding of the antisense oligonucleotide to a unique gene sequence. Antisense DNA is the non-coding strand complementary to the coding strand in double-stranded DNA. The antisense strand serves as the template for messenger RNA (mRNA) synthesis. Antisense RNA is the non-coding strand complementary to a coding sequence of mRNA, a molecule involved in translating genetic instructions into proteins. Antisense RNA hybridizes with and inactivates mRNA.

Antisense is related to a nucleotide sequence that is complementary to a sequence of messenger RNA. When antisense DNA or RNA is added to a cell, it binds to a specific messenger RNA molecule and inactivates it. Antisense is related to a gene that is derived from RNA or complementary DNA, is inserted in reverse orientation into a strand of DNA, and is used in genetic engineering to regulate genetic expression of a trait.

Mechanism

In principle, antisense technology is supposed to prevent protein production from a targeted gene. The exact mechanism by which this occurs remains uncertain. Proposed mechanisms include triplex formation, blocking RNA splicing, preventing transport of the mRNA antisense complex into the cytoplasm, increasing RNA degradation, or blocking the initiation of translation.

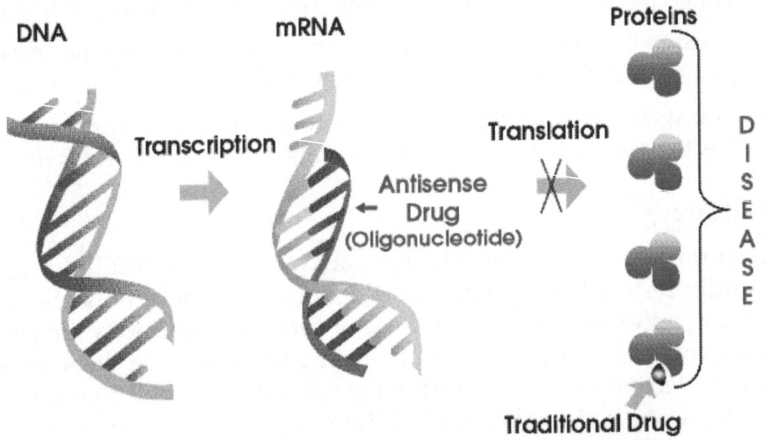

FIGURE 13-1: Mechanism of Antisense Technology.

Since mRNA represents the *'sense'* sequence, the oligonucleotide complementary to the mRNA was called *'antisense'* oligonucleotide. In any gene, the DNA strand oriented in 3′-5′ direction is called the *antisense strand.* The mRNA *base sequence, therefore, is complementary to that of the antisense strand.* The sense strand of *DNA* oriented in 5-3 direction is naturally complementary to the antisense strand of the gene. Therefore, *the base sequence of sense strand of a gene is the same as that of the mRNA produced by it* (except for T in the place of U). Hence, the mRNA produced by a gene in normal orientation is also known as *sense RNA.* An antisense gene is produced by inverting, *i.e.*, reversing the orientation of the protein encoding region of a gene in relation to its promoter. As a result, the natural sense strand of the gene becomes oriented in the 3′-5′ direction with reference to its promoter, and is transcribed.

The RNA produced by this gene has the same sequence as the antisense strand of the normal gene (except for T in DNA in the place of U in RNA), and is, therefore, known as ***antisense RNA***, or sometimes, ***asRNA***.

When an antisense gene is present in the same nucleus as the normal endogenous gene, transcription of the two genes yields antisense and sense RNA transcripts, respectively. Since the sense and the antisense RNAs are complementary to each other, they would pair to produce double stranded RNA molecules. This event makes (1) the mRNA available for translation. At the same time, (2) the double -stranded RNA molecules are attacked and degraded by double -stranded RNA specific RNases. Finally, (3) these events somehow lead to the methylation of the promoter and coding regions of the normal gene resulting in silencing of the endogenous gene.

Flavr Savr Tomato

The application of antisense RNA technology is explained using the example of tomato. In tomato, enzyme polygalacturonase (PG) degrades pectin, which is the major component of fruit cell wall. This leads to the softening of fruits and a deterioration in fruit quality. Transgenic tomatoes have been produced, which contain antisense construct of the gene encoding polygalacturonase. These transgenics show a drastically reduced expression of PG enzyme and markedly slower fruit softening.

Development

The FlavrSavr' tomatoes were first developed by Calgene company of USA using antisense RNA against polygalacturonase (PG) enzyme encoding gene. On the basis of PG gene sequence, an antisense PG gene was constructed and tomato plants were transformed. Such tomatoes were approved for marketing in USA in 1994 under the name 'FlavrSavr'. This was the first approval of transgenic food product for commercial cultivation. The main features of FlavrSavr tomatoes are listed as follows:

1. **Developed by:** These tomatoes were first developed by Calgene company of USA in 1994. Calgene has given the brand name Mac Gregor to its transgenic tomato.
2. **Shelf Life:** These tomatoes have delayed ripening in addition to the enhanced shelf-life. These tomatoes have about two week longer shelf-life than normal tomatoes without softening. In other words, they can stay on the market shelf for approximately two weeks longer without softening.
3. **Quality:** The FlavrSavr tomatoes have improved flavor and total soluble solids (TSS). These desirable features are the consequences of the fact that the fruits of FlavrSavr can be allowed to ripen on the plants. As a result, the fruits accumulate more sugars and organic acids that improve their taste and TSS content.

Mechanism Involved

These transgenic tomato plants produced both sense and antisense m-RNA for the PG gene resulting in RNA-RNA pairing. This would result in non production of PG gene product, and thus preventing the attack of PG gene upon pectin in the cell wall of ripening fruit thereby preventing softening of fruit. The antisense RNA technology has been used to suppress ethylene biosynthesis in tomato. The antisense construct of ACC synthase gene was transferred in tomato; this reduces ethylene production to <1 per cent of the normal.

The fruits remain green and hard on the plants, so that they accumulate more sugars and organic acids that make the fruits sweeter and richer in taste. After the harvest, the fruits are ripened in the warehouse by exposing them to ethylene. Another transgenic tomato variety called 'Endless Summer' was developed. The ripened fruits of this variety stay fleshy and fresh for about four weeks, which is about two weeks more than the FlavrSavr transgenic tomatoes.

Applications

Antisense technology was first effectively used in plants to alter the levels of various degradative enzymes or plant pigments. Antisense has applications in agricultural biotechnology, where it has been used to deactivate the gene that causes softening in tomatoes, and in medicine, especially in cancer and antiviral therapy. Since antisense technology focuses on preventing gene expression, it has been most widely applied to cancer gene therapy.

SUMMARY

The male sterility which is induced by the technique of genetic engineering is referred to as transgenic male sterility. The gene *barnase* is the first transgene that was used to produce male sterility by Mariani *et al.*, in 1990.

Transgenic male sterility comes under genetic male sterility and is heritable. This type of sterility has been induced in crop plants like tobacco, rapeseed, maize, *etc.* The main features of transgenic male sterility related to source of gene, types of genes involved, gene integration, nature of gene, material included, maintenance and uses are briefly discussed as follows: Identification, advantages and disadvantages of transgenic male sterility have been discussed.

Antisense refers to short DNA or RNA sequences, termed oligonucleotides, which are complementary to a specific gene sequence. The goal is to alter specific gene expression resulting from the binding of the antisense oligonucleotide to a unique gene sequence. Antisense DNA is the non-coding strand complementary to the coding strand in double-stranded DNA. The antisense strand serves as the template for messenger RNA (mRNA) synthesis. Antisense RNA is the non-coding strand complementary to a coding sequence of mRNA, a molecule involved in translating genetic instructions into proteins. Antisense RNA hybridizes with and inactivates mRNA.

Antisense technology was first effectively used in plants to alter the levels of various degrading enzymes or plant pigments. Antisense has applications in agricultural biotechnology, where it has been used to deactivate the gene that causes softening in tomatoes, and in medicine, especially in cancer and antiviral therapy.

QUESTIONS

1. **What is transgenic male sterility? What are its advantages and limitations?**
2. **Discuss briefly differences between conventional genetic male sterility and transgenic male sterility.**
3. **Define antisense gene and describe it application in plant breeding with suitable example.**
4. **Give a brief account of antisense gene technology in relation to crop improvement.**
5. **Write short notes on the following**

   (i) Flavr-Savr' tomatoes (ii) Endless Summer Tomatoes
6. **Write short notes on the following:**

   (i) Barnase gene (ii) Barstar gene

   (iii) Bar gene (iv) Antisense gene

Chapter 14

Mapping Populations

INTRODUCTION

In molecular plant breeding and genetics various types of plant materials that are used for gene mapping or gene tagging or for construction of genetic linkage maps are referred to as mapping populations. The most commonly used mapping populations include: recombinant inbred lines [RILs], near isogenic lines[NILs}, F_2 populations, backcrossed populations, bulk segregant analysis and doubled haploids. These are briefly discussed as follows:

Recombinant Inbred Lines [RILS]

Inbred lines which are developed from a cross between two pure lines or inbred lines are known as recombinant inbred lines. RILs are obtained by continuous selfing of F_2 individuals. Generally the single seed descent [SSD] method is used for developing recombinant inbred lines. RILs have been developed in several crops such as rice, soybean, capsicum, tomato, barley and cotton. Main features of RILs are briefly presented as follows:

(i) RILs are developed by successive self-pollination of F_2 individuals.

(ii) Single seed descent, pedigree and backcross methods are used for developing RILs.

(iii) RILs combine alleles from two different inbred lines or pure lines, hence the name is recombinant inbred lines.

(iv) The homozygosity is attained due to successive selfing for several generations.

(v) RILs are used for mapping of major genes as well as QTLs.

(vi) RILs are also used for construction of linkage maps or genetic maps.

(vii) When self-pollination is impossible, RILs can be constructed from full-sib crosses.

Advantages

(i) For gene mapping using RIL population, there is no need of following a particular trait.

(ii) It provides unlimited mapping possibilities being eternal.

(iii) Generally, SSR markers are used for gene mapping in RILs.

(iv) RILs are powerful tools for mapping QTLs.

Limitations

1. Development of RILs takes long time than development of F_2 population.
2. It is more expensive to develop RILs than F_2 population.

Near Isogenic Lines [NILs]

Lines with single locus difference are known as isogenic lines or isolines. Near isogenic lines refer to a group of lines which are genetically identical except for one or few loci. NILs have been developed in several crops such as wheat, oat, rice, sorghum, maize, cotton, *etc.* Main features of RILs are briefly presented as follows:

(i) NILs are developed by backcross method. Selfing is done after five backcrosses.

(ii) NILs differ from each other for one or few genes only.

(iii) NILs are homozygous and homogeneous.

(iv) NILs are used for mapping of major genes as well as QTLs.

(v) NILs are also used for construction of linkage maps or genetic maps.

(vi) NILs are relevant to both self and cross pollinated species.

Advantages

(i) NILs are used for introgressing genes from wild species or wild relatives or germplasm.

(ii) NILs are used particularly for mapping of major genes.

(iii) Generally, AFLP markers are used for gene mapping.

(iv) NILs provide unlimited possibilities for gene mapping.

Limitations

(i) Development of NILs takes long time [8-10 years] than development of F_2 population.

(ii) Repeated backcrosses have to be made. Thus, it is time consuming.

(iii) The trait of interest is to be followed.

(iv) Mapping of QTLs is difficult using NILs.

F_2 Populations

The population which is obtained by selfing of F_1 plants is known as F_2 population. In other words, F_2 population is produced by selfing of F_1 plants. The main objective is to establish linkage between the trait of interest and molecular markers. The procedure consists of following steps:

(i) A cross a made between parents with contrasting traits [say mutant and normal flower] and F_1 is obtained.

(ii) The F_1 plants are selfed giving rise to F_2 population where the trait segregates

(iii) The F_1s have normal flowers and the F_2 segregates for the mutant colour [usually 3:1 ratio normal to mutant].

(iv) F_2 plants are analysed for their molecular marker contents.

(v) If F_2 population is insufficient, individual F_2 plants can be selfed and F_3 population can be obtained for analysis of molecular markers.

Advantages

(i) F_2 populations can be used for mapping of major as well as QTL loci.

(ii) F_2 population can be developed easily as compared to RILs and NILs.

(iii) Generally, AFLP markers are used for gene mapping with F_2 population.

(iv) NILs provide unlimited possibilities for gene mapping.

Limitations

(i) Large F_2 population is required to establish linkage between traits and markers.

(ii) The gene mapping with F_2 is expensive and time consuming.

(iii) The trait of interest is to be followed.

Backcross Populations

Crossing of F_1 with one of its parents is called backcross. When is crossed with its homozygous recessive, it is called test cross. In back cross method two types of parents *viz..* recipient and donor. Recipient Parent: The parent which receives the gene of interest. It is also known as recurrent parent, because it is repeatedly crossed with F_1 Donor Parent. It is used only once in the crossing program and, hence is known as non-recurrent parent. The main features of back cross populations are given below:

(i) The recipient parent is homozygous and hence is not responsible for segregation.

(ii) The F_1 is heterozygous and, hence is responsible for segregation.

(iii) There are some heterozygotes in the segregating population.

(iv) If there is co-dominance at both marker loci, it will help in detection of such genotypes whether homozygous or heterozygous.

(v) If one of the marker loci is dominant, the parental and recombinant genotypes can be identified, if the dominant allele comes from the hybrid.

Advantages

1. It is used for developing isogenic lines which are used for gene mapping.
2. It useful for mapping of major genes.

Limitations

1. It has limited scope when character is controlled by dominant marker.
2. Repeated backcrosses have to be made. Thus, it is time consuming.
3. Mapping of QTLs is difficult using NILs.

Bulk Segregant Analysis [BSA]

This is a powerful technique which is widely used for gene mapping. It became very popular within few years after its discovery. This technique compares bulk of DNA of individuals having the same genotype. For example, in segregating population, differences between a bulk of homozygous resistant [RR] and a bulk of homozygous recessive [rr] individuals are analysed. Only the markers linked to the gene exhibit differences. The co-dominant markers show difference of band position. The dominant markers show difference of presence or absence of band. The main features of bulk segregant analysis are presented as follows:

(i) For bulk segregant analysis, F_2 population of RILs or doubled haploids segregating for gene of interest are used.

(ii) DNA fingerprinting is the most convenient method for bulk segregant analysis.

(iii) Two bulk DNA samples say one for resistant genes [RR] and other for susceptible gene [rr] are developed.

(iv) The difference between individuals of these two bulks help in identification of molecular markers tightly linked with gene of interest.

(v) It is used for mapping major genes and also filling gaps in the map.

Advantages

(i) It takes lesser time in developing mapping material than developing NILs.

(ii) It is widely used for gene mapping. It has been used for restoration of fertility in rapeseed and for disease genes in lettuce and beans.

(iii) In this technique, RFLP or RAPD markers are used for gene mapping.

Limitations

(i) This technique is rarely used for QTL mapping.

(ii) This technique requires double the markers than RILs or double haploids.

(iii) Markers which are in the repulsion phase with the gene of interest cannot be detected.

Doubled Haploids

Doubled haploids are individuals or plants which are produced from haploids through chromosome doubling. Colchicine is used for doubling of chromosomes. Thus doubled haploids have two sets of chromosomes identical to the haploid plant. Main features of doubled haploids are given as follows:

(i) Doubled haploids are produced from haploids.

(ii) They have two identical sets of haploid parent line.

(iii) They are used for rapid development of pure lines and fixation of heterosis.

(iv) They are useful in identification of mutant allele in a population.

(v) They are useful in gene mapping or construction of genetic maps.

(vi) The F_1 is heterozygous and, hence is responsible for segregation.

Advantages

1. It is used for rapid development of pure lines.
2. It is also used for gene mapping.

Limitations

1. Production of haploids is difficult.
2. This method is expensive as compared to NILs and RILs.

SUMMARY

In molecular plant breeding and genetics various types of plant materials that are used for gene mapping or gene tagging or for construction of genetic linkage maps are referred to as mapping populations. The most commonly used mapping populations include: recombinant inbred lines [RILs], near isogenic lines[NILs}, F_2 populations, backcrossed populations, bulk segregant analysis and doubled haploids.

Inbred lines which are developed from a cross between two pure lines or inbred lines are known as recombinant inbred lines. RILs are obtained by continuous selfing of F_2 individuals. Generally the single seed descent [SSD] method is used for developing recombinant inbred lines. RILs have been developed in several crops such as rice, soybean, capsicum, tomato, barley and cotton.

The population which is obtained by selfing of F_1 plants is known as F_2 population. In other words, F_2 population is produced by selfing of F_1 plants. The main objective is to establish linkage between the trait of interest and molecular markers.

Crossing of F_1 with one of its parents is called backcross. When is crossed with its homozygous recessive, it is called test cross. In back cross method two types of parents *viz..* recipient and donor. Recipient Parent: The parent which receives the gene of interest. It is also known as recurrent parent, because it is repeatedly crossed with F_1.Donor Parent. It is used only once in the crossing program and, hence is known as non-recurrent parent.

This is a powerful technique which is widely used for gene mapping. It became very popular within few years after its discovery. This technique compares bulk of DNA of individuals having the same genotype. For example, in segregating population, differences between a bulk of homozygous resistant [RR] and a bulk of homozygous recessive [rr] individuals are analysed. Only the markers linked to the gene exhibit differences. The co-dominant markers show difference of band position. The dominant markers show difference of presence or absence of band.

Doubled haploids are individuals or plants which are produced from haploids through chromosome doubling. Colchicine is used for doubling of chromosomes. Thus doubled haploids have two sets of chromosomes identical to the haploid plant.

QUESTIONS

1. **Define mapping population. Describe main features, merits and demerits of recombinant inbred lines.**
2. **Describe main features, merits and demerits of near isogenic lines.**
3. **Explain briefly main features, merits and demerits of F_2 population for gene mapping.**
4. **Discuss briefly main features, merits and demerits of back cross derived inbred lines and their use in gene mapping.**
5. **Discuss briefly main features, merits and demerits of bulk segregant analysis in relation to gene mapping.**
6. **Discuss briefly main features, merits and demerits of doubled haploid in relation to gene mapping.**
7. **Write short notes on the following:**

   (i) Gene mapping (ii) Mapping population

   (iii) Recombinant inbred lines (iv) Near isogenic lines
8. **Give a brief comparison of Recombinant inbred lines and Near isogenic lines in relation to gene mapping.**
9. **Compare Bulk segregant analysis and Doubled haploids in relation to gene mapping.**
10. **Define the following terms:**

    (i) Pure lines (ii) Isogenic lines

    (iii) Test cross iv) F_2 population

Chapter 15

Introduction to Genomics

INTRODUCTION

The term genomics was first used by Thomas Roderick in 1986. It refers to the study of structure and function of entire genome of a living organism. Genome refers to the basic set of chromosomes. In a genome, each type of chromosome is represented only once. Now genomics is being developed as a sub discipline of genetics which is devoted to the mapping, sequencing and functional analysis of genomes. Main points related to genomics are listed below.

1. It is a computer aided study of structure and function of entire genome of an organism.
2. It deals with mapping and sequencing of genes on the chromosomes.
3. It is a rapid and accurate method of gene mapping. It is more accurate than recombination mapping and deletion mapping techniques.
4. The genomic techniques are highly powerful, efficient and effective in solving complex genetic problems.
5. Now use of genomic techniques has become indispensible in plant breeding and genetics.

Types of Genomics

The discipline of genomics consists of two parts, *viz..* structural genomics and functional genomics. These are defined as under.

1. **Structural Genomics**: It deals with the study of the structure of entire genome of a living organism. In other words, it deals with the study of the genetic structure of each chromosome of the genome. It determines the size of the genome of a species in mega-bases [Mb] and also the genes present in the entire genome of a species.

2. **Functional Genomics:** The study of function of all genes present in the entire genome is known as functional genomics. It deals with transcriptome and proteome. The transcriptome refers to complete set of RNAs transcribed from a genome and proteome refers to complete set of proteins encoded by a genome.

Classification

The genomics can be classified as plant genomics, animal genomics, eukaryotic genomics and prokaryotic genomics. These are defined as follows:

(i) **Plant Genomics**: It deals with the study of structure and function of entire genome of plant species.

(ii) **Animal Genomics**: It deals with the study of structure and function of entire genome of animal species.

(iii) **Eukaryotic Genomics**: It deals with the study of structure and function of entire genome of higher [multi-cellular] organisms.

(iv) **Prokaryotic Genomics**: It deals with the study of structure and function of entire genome of unicellular organisms.

Proteomics

The study of structure and function of all proteins in an individual is called proteomics. In other words, it deals with the study of entire protein expression in an organism. Proteomics is of two types, *viz*.. (i) structural proteomics and (ii) functional proteomics. These are defined as follows:

1. **Structural Proteomics:** The study of structure of all proteins found in a living organism is called structural proteomics.
2. **Functional Proteomics:** The study of functions of all proteins found in a living organism is called functional proteomics.

In fact, proteomics is a new sub-discipline of functional genomics. It is the study of proteomes which refer to complete set of proteins encoded by a genome. A variety of techniques are now used for the study of proteomics. Now computer aided programs are available for the study of proteomics.

Metabolomics

The study of all metabolic pathways in a living organism is referred to as metabolomics. In other words, it is the computer aided information of all metabolic pathways of a living organism. Main points related to metabolomics are listed as follows:

1. It deals with the study of all metabolic pathways in a living organism.
2. It is computer aided information about metabolic pathways in a living organism.

3. It helps in identification and correction of metabolic disorders in an organism.
4. It helps in selection of individuals with normal metabolic pathways.
5. It helps in early detection of genetic disorders associated with metabolic pathways.

Genomics in Crop Plants

The discipline of genomics is of recent origin. The genome mapping was first completed in free living bacteria *Haemophillus influenza* in 1995. Subsequently, genome sequencing work was intensified both in prokaryotes and eukaryotes. In plants, genome sequencing was first completed in *Arabidopsis thaliana* (a weedy relative of mustard) followed by rice (*Oryza sativa*). Now genome sequencing work has been completed in more than 40 crop plants. The genome size and gene number in some crop plants are presented in Table 15-1.

TABLE 15-1: Genome Sequencing in some Crop Plants

Sl.No.	*Year*	*Name of Scientist(s)*	*Crop Species*	*Botanical Name*
1	2002	A Team of Scientists	Rice	*Oryza sativa*
2	2006	Tuskan, *et al.*	Poplar	*Populus tricocrpa*
3	2007	Jaillon, *et al.*	Grapes	*Vitis vinifera*
4	2007	Jaillon, *et al.*	Papaya	*Carica papaya*
5	2008	Ming, *et al.*	Apple	*Malus domesticus*
6	2008	Sato, *et al.*	Lotus	*Lotus japonicas*
7	2009	Schnable, P *et al.*	Maize	*Zea mays*
8	2009	Paterson, A *et al.*	Sorghum	*Sorghum bicolor*
9	2009	Haung *et al.*	Cucumber	*Cucumis sativa*
10	2010	Schmutz, *et al.*	Soybean	*Glycine max*
11	2010	AP Chan, *et al.*	Cartor bean	*Ricinus communis*
12	2010	Vogel, J *et al.*	Brachypodium	*Brachypodium*
13	2010	Reccardo Velasco, *et al.*	Peaches	*Prunus persica*
14	2010	Eman K Al-Dous *et al.*	Date palm	*Phoenix dactylifera*
15	2012	Angelique D'Hont *et al.*	Banana	*Musa acuminate*
16	2011	Varshney, *et al.*	Pigeon pea	*Cajanus cajan*
17	2011	PGS Consortium,	Potato	*Solanum tuberosum*
18	2011	Harm van Bakel, *et al.*	Cannabis	*Cannabis sativa*
19	2011	Bernd Weisshaar, *et al.*	Sugar beet	*Beeta vulgaris*
20	2011	Tina T. Hu *et al.*	Arabidopsis	*Arabidopsis lyrata*
21	2011	Wang, *et al.*	Mustard	*Brassica rapa*
22	2012	Zheng G *et al.*	Foxtail Millet	*Setaria italic*
23	2012	Tomato Genome Consortium.	Tomato	*Solanum lycopersicum*
24	2012	Angelique D'Hont *et al.*	Banana	*Musa acuminate*
25	2012	Garci- Mas J *et al.*	Melon	*Cucumis melo*

Sl.No.	Year	Name of Scientist(s)	Crop Species	Botanical Name
26	2012	Wang, *et al.*	Flax	*Linum usitatissimum*
27	2012	Kunbo Wang *et al.*	Cotton	*Gossypium raimondii*
28	2012	Angelique D' Hont *et al.*	Sweet Orange	*Citrus sinensis*
29	2012	Wang and Jhou.	Clementine Orange	*Citrus clementina*
30	2013	Ling *et al.*	Wheat	*Triticum urartu.*
31	2013	Genome Biology.	Tobacco	*Nicotiana sylvestris L.*
32	2013	Phytozome.	Common bean	*Phaseolus vulgaris.*
33	2014	Kim *et al.*	Pepper	*Capsicum annum.*
34	2015	CSIR Scientists	Tulsi	*Oscimum sanctum*
35	2016	Studer Anthony	Grass	*Diacanthelium.*
36	2016	Bertioli *et al.*	Wild Peanut	*Arachis ipaensis*
37	2017	Jarvis *et al.*	Quinoa	*Chenopodium quinoa.*
38	2017	Badouin *et al.*	Sunflower	*Helianthus annuus*
39	2017	Luo *et al.*	Aegilops	*Aegilops tauschii*
40	2018	Appels, *et al.*	Bread wheat	*Triticum aestivum*
41	2018	Chang *et al.*	Lablab bean	*Lablab purpureus*
42	2018	Sato *et al.*	Lotus	*Lotus japonicus*
43	2018	Giga DB	Wild Vigna	*Vigna subterranea*
44	2018	Chang *et al.*	Moringa	*Moringa oleifera*
45	2018	VanBuren *et al.*	Rubus	*Rubus accidentalis*
46	2019	Zhao *et al.*	Teak	*Tectona grandis*
47	2019	Song *et al.*	Solanum spp.	*Solanum aethiopicum*

Genome Mapping in India

The discipline of genomics is of recent origin. In India, the genome mapping work is carried out at the following centers:

1. National Research Centre for Plant Biotechnology, IARI, New Delhi.
2. International Centre for Genetic Engineering and Biotechnology, New Delhi.
3. Jawaharlal Nehru University, New Delhi.
4. National Botanical Research Institute, Lucknow.

Now functional genome projects are being launched by the Depart of Biotechnology [DBT] and ICAR. The DBT has initiated such work on several crops such as rice, wheat, maize, chick pea, banana, tomato, Brassica, *etc.* The ICAR has created genome mapping facilities for rice at NRCPB, IARI, New Delhi.

Gene to be Mapped

In genome research, both types of gene *viz.*. major genes (oligogenes) and minor genes (polygenes) can be easily mapped. The mapping of polygenic traits is possible by genome mapping techniques which is not possible by conventional gene mapping techniques such as recombination mapping and deletion mapping. Genome mapping is done for morphological, productivity, resistance, quality, agronomic and some special traits as discussed below:

1. Morphological Traits: It includes highly heritable traits such as shape, size and color of leaf, flower calyx, corolla, *etc.* It also includes surface of leaf and stem (hairiness and smoothness).
2. Productivity Traits: Such traits differ from species to species.
3. Resistance Traits: Such traits include resistance to diseases, insects, drought, soil salinity, soil alkalinity, soil acidity, heat, frost water logging, cold, *etc.*
4. Quality Traits: Such traits include nutritional quality, market quality, keeping quality and industrial quality.
5. Agronomic Traits: Such traits include earliness, plant height, plant type, *etc.*
6. Special Traits: Such traits include gene controlling male sterility, self-incompatibility, photo and thermo insensitivity, toxic substances, apomixes, adaptation, *etc.*

Some Genome Mapping Laboratories

The genome mapping work is carried out by collaborative efforts of various International and National Research Laboratories of public domain. The list of some organizations where genome sequencing and mapping work on different crops is carried out is presented in Table 15-2.

TABLE 15-2: List of some Genome Sequencing Organization

Sl. No.	*Name of Crop*	*Scientific Name*	*Name of Genome Project*	*In Collaboration with*
1	Arabidopsis	*Arabiddopsis thaliana*	Arabidopsis Genome Initiative.	USA, China and Japan.
2	Brassica	*Brassica* spp.	Brassica Genome Gate way	UK, USA, France, Germany, Australia, South Korea, Poland and China.
3	Wheat	*Tricum aestivum*	International Wheat genome Sequencing Consortium	UK Scientists from Uni. Of Liverpool, Bristol and John Innes.
4	Rice	*Oriza* spp.	International Rice Genome Sequencing Project	Mexico, USA, China and japan
5	Cotton	*Gossypium* spp.	International Cotton Genome Initiative Project	USA, France, Australia
6	Soybean	*Glycine max*	Uni. Purdue, North Carolina Uni, and 18 other centres	United States of America

Sl. No.	*Name of Crop*	*Scientific Name*	*Name of Genome Project*	*In Collaboration with*
7	Potato	*Solanum tuberosum*	Potato Genome Sequencing Consortium	USA, UK, Russia, China, India, Poland, Chile,New Zealand, Nether land, Peru, *etc.*
8	Tomato	*Lycopersicum esculentum*	Tomato Genome Consortium	China, japan, Germany,Korea, Spain, UK, USA, Israel, Netherland and Others
9	Pigeon pea	*Cajanus cajan*	International Initiative for Pigeonpea Genomics	India (ICRISAT), China, USA, Mexico, *etc.*
10	Apple	*Malus domesticus*	Apple Genome sequencing Initiative	Washington State Uni. USA and NRI COMPETITIVE GRANT

Applications in Crop Improvement

Genomics has various practical applications in crop improvement. The genome mapping provides information about genome size, gene number, gene mapping, gene sequencing, evolution of crop plants, gene cloning, identification of DNA markers, marker assisted selection and construction of linkage maps, *etc.*

1. **Genome Size:** Genome mapping is a very useful technique for determining genome size in various plant species. In the plant species studied so far, the largest genome size has been reported in maize (25000 Mb) and the smallest in *Arabidopsis thaliana* (120MB).
2. **Gene Number:** Genome mapping provides information about gene number in a species. In crop plants studied so far, the maximum number of genes have been reported in rice (56,000).
3. **Gene Mapping:** Genome research is very much useful in mapping/tagging of genes on the different chromosomes of the genome. In other words, it helps in large scale discovery of new genes in a genome.
4. **Gene Sequencing**: Genome mapping helps in determining the order of genes on the chromosomes. The order of genes is determined on each chromosome of a genome.
5. **Evolution:** Genome mapping provides information about the evolution of different species. It measures the association between different genomes and thus provides information about the relatedness or evolutionary biology of crop plants.
6. **Gene Cloning**: Genome research is very much useful in making multiple copies of a gene and transfer of the same from one genotype to another. Thus, it aids in specific gene transfer.
7. **Identification of DNA Markers**: Genome mapping techniques are useful in identification of DNA markers which can be used in molecular breeding *i.e.* in marker assisted selection. The mapping populations developed from

inter-specific crosses have high level of polymorphism for DNA markers than those mapping populations derived from intra-specific crosses.

8. **Marker Assisted Selection**: Var5ious DNA markers such as RFLP, AFLP, RAPD, SSR, *etc.* are used for marker assisted selection. The effect of DNA marker is correlated with morphological markers and then selection is made for a particular trait. The selection based on DNA markers is more reliable, because DNA markers are not influenced by environmental factors.
9. **Transgenic Breeding**: Genome mapping is useful in gene cloning. The gene of interest can be cloned and used for developing transgenic plants. Transgenic breeding permits direct gene transfer bypassing sexual process.
10. **Construction of Linkage Maps**: Genome mapping helps in construction of linkage groups. The linkage groups can be constructed from information of gene mapping and gene sequencing.
11. **QTL Mapping**: The genome sequencing technique is widely used for mapping of quantitative trait loci (QTL). Mapping of QTL or polygenic trait is not possible by conventional method, *viz.*. recombination mapping and deletion mapping.

Achievements

In crop plants, genome mapping work has been done in more than 40 field and fruit crops. Important crop plants in which genome sequencing has been done include Arabidopsis, Brassica rapa, Rice, Corn, Cucumber, Soybean, Pigeon pea, Tomato, Potato, Melon, Flax, Cotton from field crops; and oranges, apple, peaches, papaya, grapes, Date Palm, Banana, *etc.* from fruit crops. The genome sequencing of many more crops is in progress.

Limitations

There are some limitations of genome mapping such as high cost, high technical skill, laborious work, limited genes, and lack of proper markers. These are briefly discussed as follows:

1. **Costly Method**: Genome research requires well equipped sophisticated laboratory with costly chemicals and glassware. Thus lot of funds are required for carrying genomic research. Lack of adequate funds sometimes becomes limiting factor in the progress of such project.
2. **High Technical Skill**: Genome mapping work requires high technical skill. It requires training of scientists in the specialized field of genomics. It also requires International collaboration with other leading genome research laboratories, which sometimes becomes a limiting factor. The International collaboration is possible when the crop on which genome research work is to be carried out is of global significance.
3. **Laborious Work:** Genome mapping requires detection of various DNA markers [AFLP, RFLP,RAPD, SNP, SSR, *etc.*] which is a laborious and

time consuming work. Huge populations related to F2, RILs, NILs and doubled haploids need to be screened for such purpose. This limits the progress of the work.

4. **Limited Genes**: Firstly, limited numbers of genes are available for development of transgenics. Secondly, such genes are protected under Intellectual Property Rights and, therefore, cannot be used for developing transgenic plants.
5. **Lack of Proper Markers**: Most of the useful agronomic traits are governed by polygenes and are complex in nature. Tightly linked DNA markers are yet to be identified for such characters.

FUTURE THRUSTS

Considerable research work on genome sequencing and mapping has been done on various crop plants so far. In future, for rapid progress of genome sequencing work important points which need consideration include funding, training, material sharing, research priorities, important traits, selection of species, *etc.* These are briefly discussed as follows:

1. **Funding**: Since genome mapping projects are very expensive, there is need of International collaboration for supporting such prestigious projects.
2. **Training**: The laboratories which are well equipped should impart training to scientists from various countries for human resource development in genome research.
3. **Material Sharing**: The leading laboratories should develop and distribute frame work DNA markers to other research laboratories for their use and further research.
4. **Research Priorities:** For getting International collaboration, areas of common interest or global significance should be identified for genomic research work.
5. **Important Traits**: In genome mapping, the major emphasis needs to be given to characters of economic importance such as productivity, quality and resistance to biotic and abiotic stresses. The experience gained in genome mapping of one species may be helpful in the study of related species.

SUMMARY

The term genomics was first used by Thomas Roderick in 1986. It refers to the study of structure and function of entire genome of a living organism. Genome refers to the basic set of chromosomes.

The discipline of genomics consists of two parts, *viz..* structural genomics and functional genomics. Structural Genomics deals with the study of the structure of entire genome of a living organism. Functional Genomics is the study of function of all genes present in the entire genome is known as functional genomics. It deals with transcriptome and proteome. The transcriptome refers to complete set of

RNAs transcribed from a genome and proteome refers to complete set of proteins encoded by a genome.

Genomics has various practical applications in crop improvement. The genome mapping provides information about genome size, gene number, gene mapping, gene sequencing, evolution of crop plants, gene cloning, identification of DNA markers, marker assisted selection, construction of linkage maps, *etc.*

There are some limitations of genome mapping such as high cost, high technical skill, laborious work, limited genes, and lack of proper markers. These are briefly discussed.

In future, for rapid progress of genome sequencing work important points which need consideration include funding, training, material sharing, research priorities, important traits, selection of species, *etc.*

QUESTIONS

1. **Define genomics. Describe in brief structural and functional genomics.**
2. **Define proteomics. Describe in brief structural and functional proteomics.**
3. **Discuss briefly advantages of bioinformatics.**
4. **Describe briefly plant genome sequencing. Give estimated genome size in different plant species available so far.**
5. **Discuss briefly various applications of genomics in crop improvement.**
6. **Explain briefly various limitations of genomics.**
7. **Describe in brief future prospects of genomic research related to field crops.**
8. **Write short notes on the following:**
   (i) Genomics (ii) Proteomics
   (iii) Structural genomics (iv) Functional genomics
9. **Explain the following terms:**
   (i) Metabolomics (ii) Transcriptome
   (iii) Proteome

Introduction to Bioinformatics

INTRODUCTION

Bioinformatics refers to the computer aided study of biology particularly genetics and molecular biology. Now the science of bioinformatics is gaining increasing importance in life science specially in the field of molecular biology and plant genetic resources. Main points related to bioinformatics are listed as under.

1. It is the application of information technology in the study of biology. Thus it utilizes information science for the study of biology.
2. It is used for computer based analysis of large scale data derived from genome sequencing.
3. It is used for analysis of data related to genomics, proteomics, metabolomics and other biological aspects.
4. It has wide applications in handling data related to plant genetic resources.

Types of Bioinformatics

Bioinformatics is broadly divided in to two groups, *viz*.. animal bioinformatics and plant bioinformatics. These are defined as follows:

1. **Animal Bioinformatics:** It deals with computer aided study of genomics, proteomics and metabolomics in animal species.
2. **Plant Bioinformatics**: It deals with computer aided study of to genomics, proteomics and metabolomics in plant species.

Applications in Plant Breeding

Bioinformatics or computer aided programs have wide practical applications in genetics and plant breeding. Some important applications of bioinformatics in plant breeding and genetics include (1) varietal information system, (2) plant genetic resources database, (3) biometrical analysis, (4) storage and retrieval of data, (5) studies on plant modeling, (6) pedigree analysis, (7) preparation of reports, (8) updating of information, (9) diagrammatic representation, and (10) planning of breeding programs. These are briefly discussed as follows:

1. Varietal Information System

Bioinformatics has useful application in developing varietal information system. In connection with plant variety protection (PVP) Act, various terms such as extant variety, candidate variety, reference variety, example variety and farmer's variety are frequently used. Hence, knowledge of these terms is essential. These are defined below:

Extant Variety: All the released, notified and unprotected varieties.

Candidate variety: A variety to be registered under Plant Variety Protection Act is referred to as candidate variety.

Reference Variety: All released and notified extant varieties of common knowledge which are in seed production chain.

Example Variety: A variety that is used for comparison for a particular character is called example variety.

Farmers' Variety: A variety that has been developed by a farmer and used for commercial cultivation for several years is called farmers variety.

The detailed information about various types of varieties can be developed using highly heritable characters. Such information can be used in various ways as given below.

(i) For varietal identification in DUS testing.

(ii) In grouping of varieties on the basis of various highly heritable characters.

(iii) In sorting out of cultivars for use in pre-breeding and traditional breeding.

The information can be stored in the computer memory and be retrieved as and when required.

2. PGR Database

The genetic material of plants which is of value for present and future generations of people is referred to as plant genetic resources. It is also known as gene pool, genetic stock and germplasm. The germplasm is evaluated for several characters such as highly heritable morphological, yield contributing characters, quality characters, resistance to biotic and abiotic stresses and characters of agronomic value.

International Plant Genetic Resources Institute (IPGRI), Rome Italy has developed descriptors and descriptor states for various crop plants. Such descriptors help in uniform recording of observations on germplasm of crop plants all over the world. Thus huge data is collected on crop genetic resources for several years. Bioinformatics plays an important role in systematic management of this huge data. Bioinformatics is useful in handling such data in several ways as follows:

(i) It maintains the data of several locations and several years in systematic way.

(ii) It permits addition, deletion and updating of information.

(iii) It helps in storage and retrieval of huge data.

(iv) It also helps in classification of PGR databased on various criteria.

(v) It helps in retrieval of data belonging specific group such as early maturity, late maturity, dwarf types, tall types, resistant to biotic stresses, resistance to abiotic stresses, superior quality, marker genes, *etc.*

All such data can be easily managed by computer aided programs and can be manipulated to get meaningful results.

3. Biometrical Analysis

In crop improvement, various biometrical analyses are performed. Important biometrical analyses that are performed in plant breeding and genetics are given below.

(i) Simple measures of variability such as mean, standard deviation, standard error, coefficient of variation, *etc.*

(ii) Correlations: It includes genotypic, phenotypic and environmental correlations. It also includes simple, partial and multiple correlations.

(iii) Path Coefficients: It includes analysis of genotypic, phenotypic and environmental paths.

(iv) Discriminant function analysis.

(v) Metroglyph analysis and D^2 statistics

(vi) Stability analysis

(vii) Diallel, partial diallel, line x tester triallel, quadriallel, biparental and triple test cross analysis.

(viii) Generation mean analysis, *etc.*

All these analyses can be easily performed through computer aided programs.

4. Storage and Retrieval of Data

In crop improvement, huge data is collected on the following aspects:

(i) Segregating populations: Single plant selections are made in segregating populations and data are recorded on various characters such as yield components, quality characters, resistance to biotic and abiotic stresses, *etc.*

(ii) Multi-location Experiments: Such experiments are conducted mainly for identification and release of new varieties and hybrids and also for assessment of varietal stability.

(iii) **Multi-seasonal Experiments:** Such experiments are conducted for several years (3-5 years) for identification of new varieties and hybrids.

The above data remain in active use generally for two decades. Handling of such a huge data is a difficult task. However, such data can be easily stored in various storage devices such as hard disks, compact disks, pen drive, data cards, *etc.* Storage of data in computers require less space and is very safe as compared to storage of data in paper registers and files.

5. Studies on Plant Modeling

Computers are useful tools for undertaking studies on modeling of plants. First the theoretical model can be prepared with the help of computer keeping in view various plant characters. Then such model plants can be developed through hybridization and directional selection. This types of studies are useful in developing crop ideotypes or ideal plant types in different field crops. First the conceptual model is prepared and then efforts are made to achieve such model by combining desirable genes from different sources into a single genotype through appropriate breeding procedures. Such studies have been made in field pea.

6. Pedigree Analysis

Computer aided studies are useful in pedigree analysis of various cultivars and hybrids. Information about the parentage of cultivars and hybrids is entered into the computer memory which can be retrieved any time. The list of parents which are common in the pedigree of various cultivars and hybrids can be sorted out easily. It helps in the pedigree analysis which in turn can be used in planning plant breeding programs especially in the selection of parents for use in hybridization programs. The study of proteomics also helps in pedigree analysis.

7. Preparation of Reports

After biometrical analysis of data, results are interpreted and various types of reports or documents are prepared. In crop improvement, following types of reports are prepared.

(i) Research Project Report: The annual progress report of each project is prepared and salient findings are documented.

(ii) Monthly, quarterly, half yearly and annual progress reports of all the research projects are also prepared.

(iii) Sometimes, bulletin and booklets are prepared to document specific information for adoption and benefit of farmers.

(iv) Research papers and popular articles are prepared based on research findings.

(v) Germplasm catalogues are prepared for various characters.

Such reports can be easily be prepared with the help of computers using MS Word program. This information can be stored in computer memory and reused as and when required. The editing and updating of reports can be done any time without much extra efforts.

8. Updating of Information

In plant breeding and genetics, results of multi-seasonal and long term experiments require continuous updating. Computers have made this task very simple. The information related to any experiment which is already stored in the computer memory, can be updated any time by editing the concerned file. Any portion of information can be deleted or revised easily.

9. Diagrammatic Representation

Inclusion of diagrams makes the reports, research papers, articles, bulletins, *etc.* more attractive, informative and easily understandable. The following types of diagrams are made in plant breeding.

(i) Line diagrams, bar diagrams, histograms and pie diagrams.

(ii) Cluster diagram: It is prepared when data is subjected to D2 analysis.

(iii) Path diagram: It is prepared when data is subjected to path coefficient analysis.

(iv) Vr-Wr Graph: It is prepared when data is subjected to Hayman's graphical approach of diallel cross analysis.

(v) Metroglyph Chart: It is prepared when data is subjected to Metroglyph analysis.

All these diagrams can be easily prepared with the help of computer using specific program.

10. Planing to Breeding Programs

Plant breeders have to plan various breeding programs every year. Computers are useful tools in such planning. The following activities can be easily olanned with the help of computers.

(i) Sowing plans of various breeding experiments.

(ii) Selfing and crossing plans.

(iv) Breeder seed production plan.

(v) Hybrid seed production plan.

(vi) Germplasm collection, conservation, evaluation, distribution, utilization and documentation plan.

(vii) Screening plan of breeding material against biotic and abiotic stresses.

(viii) Selection, quality evaluation and multi-location testing plans.

All the above plans can be easily prepared with the help of computer well in advance. This is very important for proper implementation of various breeding programs. Computers are also useful in printing out labels and list of observations to be recorded in various breeding experiments.

Advantages

Bioinformatics has several practical applications in genetics and plant breeding as discussed above. There are several advantages of bioinformatics such as systematic approach, study of evolution, rapid method, identifying similar genes, accuracy, designing of primers and handling of huge data. These are discussed as follows:

1. **Systematic Approach:** It provides systematic information about genetics, proteomics and metabolomics of living organisms. This information is useful in planning various breeding and genetic programs.
2. **Study of Evolution**: It helps in finding evolutionary relationship between two species. Studies of nucleotide and protein sequences help in such matter. The closely related organisms have similar sequences and distantly related organisms have dissimilar sequences. The time of divergence between two species can also be estimated from such studies. Thus bioinformatics helps in the study of evolutionary biology. It helps in drawing phylogenetic trees [trees of relatedness].
3. **Rapid Method:** It is a rapid method of gene mapping and sequencing. Earlier methods of gene mapping were time consuming and pains taking. Bioinformatics has made this task very simple. Now gene hunting has become faster, cheaper and systematic.
4. **Identification of Similar Genes**: Computer aided studies help in identification of similar genes in two species. For example, genes similar for biotic and abiotic stresses in two species can be easily detected. Computer aided programs has made it possible to group proteins into families based on their relatedness.
5. **Accuracy:** The computer aided information has very high level of accuracy and is highly reliable.
6. **Designing Primers:** Computer aided programs are useful in designing primers for PCR. Such primers are used to sequence unknown genes or gene of interest.
7. **Handling Huge Data:** In life science, computer aided programs are useful in storing, organizing and indexing huge database.

Limitations

Computer aided programs have helped in better understanding of various processes of life science. However, there are some limitations of bioinformatics such as high initial cost, training, regular power supply, anti-virus vaccine and maintenance. These are briefly discussed as follows:

1. **High Initial Cost:** Bioinformatics requires sophisticated laboratory of molecular biology for in-depth study of biomolecules. Establishment of such laboratory requires lot of funds.

2. **Requires Training:** Computer based study of life science requires some training about various computer programs applicable for the study biomolecules. Thus special training is required for handling of computer based biological data.

3. **Regular Power Supply**: There should uninterrupted power supply for computer aided biological investigations. Interruption of power may sometimes lead to loss of huge data from computer memory.

4. **Anti-Virus Program:** There should be a antivirus program to keep computer virus free. Viruses pose several problems such as deletion of data and corruption of the program.

5. **Costly Maintenance**: Maintenance and upkeep of molecular laboratories involves lot of expenditure which sometimes becomes a limiting factor for computer based molecular studies.

SUMMARY

Bioinformatics is the computer aided study of biology particularly genetics and molecular biology. Now the science of bioinformatics is gaining increasing importance in life science especially in the field of molecular biology and plant genetic resources. Bioinformatics is broadly divided in to two groups, *viz*.. animal bioinformatics and plant bioinformatics. Animal Bioinformatics deals with computer aided study of genomics, proteomics and metabolomics in animal species. Plant Bioinformatics deals with computer aided study of genomics, proteomics and metabolomics in plant species.

Bioinformatics or computer aided programs have wide practical applications in genetics and plant breeding. Some important applications of bioinformatics in plant breeding and genetics include (i) varietal information system, (ii) plant genetic resources database, (iii) biometrical analysis, (iv) storage and retrieval of data, (v) studies on plant modeling, (vi) pedigree analysis, (vii) preparation of reports, (viii) updating of information, (ix) diagrammatic representation, and (x) planning of breeding programs. These are briefly discussed.

Bioinformatics has several practical applications in genetics and plant breeding as discussed above. There are several advantages of bioinformatics such as systematic approach, study of evolution, rapid method, identification of similar genes, accuracy, designing of primers and handling of huge data.

Computer aided programs have helped in better understanding of various processes of life science. However, there are some limitations of bioinformatics such as high initial cost, training, regular power supply, anti-virus vaccine and maintenance. These are briefly discussed.

QUESTIONS

1. **Define bioinformatics and discuss its applications in plant breeding.**

2. **Explain briefly the role of computers in handling PGR data in plant breeding.**

3. **Discuss role of computers in (i) pedigree analysis and (ii) updating of information**

4. **Describe briefly role of computers in (i) biometrical analysis and (ii) report writing.**

5. **Explain briefly the role of computers in (i) plant modeling and (ii) diagrammatic representation.**

6. **Describe briefly the role of computers in (i) maintaining varietal information system and (ii) storage and retrieval of data.**

7. **Describe advantages and limitations of bioinformatics in crop improvement.**

8. **Write short notes on the following:**

   (i) Varietal information system
   (ii) PGR Database
   (iii) Biometrical analysis
   (iv) Plant modeling

9. **Write short notes on the following:**

   (i) Extant Variety
   (ii) Candidate Variety
   (iii) Reference Variety
   (iv) Example Variety

Chapter 17

Gene Revolution

INTRODUCTION

Gene revolution refers to quantum jump in the productivity of various field crops through use of transgenic cultivars. It has been achieved through the application of Agricultural Biotechnology. Hence agricultural biotechnology is referred to as the 'Evergreen Revolution' or the 'Gene Revolution. Modern biotechnology plays a vital role in the"Gene Revolution". Main points related to gene revolution are listed below.

1. Gene revolution is also known as ever green revolution.
2. Gene revolution has been achieved through the application of agricultural biotechnology.
3. It has been achieved by private seed companies.
4. It is a privately owned technology.
5. It has been funded by private seed companies.
6. Production and marketing of GM crops is controlled by private seed companies.
7. Private companies do not disclose the parentage of the hybrids.

Main Features of Gene Revolution

Presently, the genetically modified crops [GM crops] are gaining increasing importance the world over. The GM crops are expected to spread around the world and lead to a global Gene Revolution. The main features of gene revolution or ever green revolution in terms of ownership, funding source, accessibility of the technology, application of IPR, gene transfer, method employed, transparency, benefits, bio-safety, bioethical measures, *etc.* in relation to green revolution are briefly discussed hereunder.

1. **Ownership of Technology**: It is privately owned technology. The ownership of the gene revolution is vested in the Private seed Companies or Corporations. Genetically modified crops are largely the product of private industry. This is partly because new technologies are far more costly than existing ones, and the biotechnology industry was able to gather the necessary funds to develop these technologies long before public awareness of GM crops could lead to publicly generated funding for GM crop development. Successful companies typically focus on their markets with the intent of generating profit. With regard to agricultural biotechnology, companies in the United States and elsewhere have thus far created primarily seeds that farmers in developed countries can and will purchase: corn and soybeans that can tolerate a particular herbicide, corn and cotton that are resistant to particular pests, and food crops that last longer on the supermarket shelf.
2. **Funding Agencies:** It is funded by private seed companies. Six seed companies *viz..*, Monsanto, Syngenta, Bayer Crop Science, DuPont, Dow and BASF Plant Science control practically the entire research and output in the field of transgenic plants. In the beginning, the green revolution program was funded by the Rockefeller Foundation. Later on other organizations such as the World Bank, Food and Agricultural Organization of the United Nations (FAO), United States Agency for International Development (USAID), and other national and international organizations joined the Rockefeller Foundation to make this program successful. It succeeded in terms of increasing food production in Asia, Latin America, and even parts of the industrialized world such as Great Britain.
3. **Accessibility of Technology:** The transgenic seed belongs to various seed companies, which strictly control its production and marketing. Developed countries grow such costly seeds because of their high yield under their environmental conditions. Thus, in developed nations, GM crop technology could revolutionize farming. However, the current GM crop seed varieties are neither affordable nor useful to most of the poorer farmers in the world; hence, their revolutionary impact in the developing world has been limited so far.
4. **Application of Intellectual Property Rights**: Intellectual Property Rights are applicable to the products of transgenic technology *i.e.* patenting is done. A fundamental challenge in this newest agricultural movement that did not arise during the Green Revolution is the definition and treatment of intellectual property (IP). IP issues are central to the Gene Revolution that can often limit the valuable diffusion of such technology.
5. **Gene Transfer:** The gene transfer is possible between two totally unrelated organisms or across any species. The gene transfer is possible from animals to plants and vice versa. The Gene Revolution combined with genetic engineering, allows unique trait combinations across species. For example, daffodil and bacterial genes can be introduced into the rice genome so that the rice produces beta-carotene, the precursor of vitamin A.

6. **Methods Involved:** Highly sophisticated biotechnological approaches are used for development of transgenic plants. The science and technology required to create GM crop seeds are far more complicated than the science and technology used to create Green Revolution agricultural advancements
7. **Transparency:** The information is kept secret by the seed companies responsible for developing the technology. GM seeds are created largely through private enterprises rather than through public-sector efforts
8. **Benefits:** The gene revolution is expected to bring even greater benefits to the small and marginal farmers than green revolution. The benefits of the current varieties of GM crops include yield increase, reduced agricultural inputs such as pesticides and fertilizers, reduced vulnerability to the vagaries of nature, and improved nutritional content. For the most part, these benefits have been limited to parts of the developed countries to which current GM crop development and marketing have been targeted and also in those countries that have allowed cultivation of such crops.
9. **Bio-safety Measures:** Bio-safety measures are required in case of genetically modified crop plants.
10. **Bio-ethical Measures:** Bio-ethical measures have to be adopted in case of genetically modified crop plants.
11. **Organic farming of end product:** Organic farming of genetically modified crop plants is not permitted.
12. **End product:** The end product of agricultural biotechnology is a transgenic cultivar.
13. **Evolution of Super Weeds:** Evolution of super weeds is possible after out crossing of transgenic plants with wild relatives of cultivated plants
14. **Technical Skill Required:** High technical skill is required in developing genetically engineered crop plants.
15. **Initial Cost of Infra Structure:** The initial cost of infra structure required for evolving cultivars through agricultural biotechnology is very high. Costly equipments, glassware and chemicals are required to carryout research in the biotech laboratory.
16. **Sexual Process for Gene transfer:** For development of initial transgenic plants, the sexual process is not involved for gene transfer.
17. **Laboratory required:** A well equipped research Laboratory is required to carryout work for development of transgenic plants.
18. **Time Requirement:** It is a rapid method of crop improvement. The development of new transgenic variety takes 4-5 years as compared to 10-12 years taken by conventional methods of plant breeding.
19. **Involvement of Transgene:** Trans gene is involved in the development of transgenic cultivars. In case of green revolution, the high yielding cultivars were developed by using dwarfing genes.

20. **Characters Improved:** In the green revolution, major emphasis has been laid in increasing the productivity, whereas in gene revolution the major emphasis has been given in the development of cultivars resistant to insects and herbicides.

Green Revolution and Gene Revolution

The Green Revolution and the gene revolution can be compared in four basic areas: science and technology, funding sources, places where the movement occurred or is occurring, and the policies and political motivations surrounding each movement.

1. Science and Technology

The Green Revolution made a considerable impact in agriculture in the developing world, and to a limited extent in developed countries. The integrated approach of breeders, people from pesticide and chemical fertilizers industries and irrigation facilities helped farmers in realizing higher crop yield. Now GM crops are being developed that survive on less water, that survive in soil heavy in salt or metals such as aluminum, that convert or "fix" nitrogen from the air, and that produce vaccines against common diseases such as cholera and hepatitis B (Byrne *et al.*, 2004). Commercial application of biotechnology has taken place primarily in the United States and primarily through the private sector.

2. Funding Sources

Initially, the Rockefeller and Ford Foundation, provided funding for the Green Revolution. The governments of Mexico, India, and several other countries also funded the green revolution program. In 1971, while the Green Revolution was bearing its first fruits in many parts of the world, the Consultative Group on International Agricultural Research (CGIAR) governing functioning of 16 International Research Centers working in more than 100 countries—was established. With the creation of CGIAR, support for developing world agriculture became more broad-based and included European nations, Canada, and Japan.

Now, some agricultural biotechnology companies are willing to work with government research institutions to develop crops that would be profitable and affordable by farmers in developing countries. In addition, they are willing to donate a substantial portion of their scientific knowledge, such as genomes of key food crops, to increase agricultural knowledge in the developing world. In this way, the challenges related to intellectual property may be lessened.

3. Places of Revolution

The Green Revolution was a success, in terms of its stated objectives, in Mexico and the rest of Latin America, India, and much of Southeast Asia. However, the Green Revolution made little significant impact in most areas of Africa. In Africa, the Green Revolution has yet to improve food production in a sustainable way.

The technologies introduced in Asia and Latin America in the Green Revolution generally required not more land, but chemical fertilizer and timely irrigation. Big farmers could take more advantage of gene revolution than small farmers. The gene revolution technologies are spreading rapidly. Moreover, farmers will adopt only those technologies that are cheaper, produce more and reduce the cost of production as compared to current technologies. It is expected that all farmers, small or large, would be able to take advantage of the Gene Revolution technology.

4. Policies and Political Motivation

Now, public concerns and national and international regulations are the driving force behind whether GM crops are adopted or rejected in various parts of the world, because wider public scrutiny and the newness of the science have led to concerns about environmental and health risks of GM crops.

Moreover, now a variety of prominent nongovernmental organizations (NGOs) that are concerned about the influence of multinational corporations, environmental degradation, crop diversification, food safety and globalization. These NGOs were very much less influential during the Green Revolution.

TABLE 17-1: Comparison of Green Revolution and Gene Revolution

Sl.No.	*Particulars*	*Green Revolution*	*Gene Revolution*
1	Definition	It refers to the quantum jump in the productivity of cereals especially wheat and rice through the use of dwarfing genes	It refers to the quantum jump in the productivity of crop plants through development of transgenic plants
2	Ownership of the technology	It is publicly owned technology. The ownership of the green revolution was vested in the farmer.	It is privately owned technology. The ownership of the gene revolution isvested in the Private seed Companies.
3	Funding Agencies	It was funded by Government. It was funded mainly by Rockefeller Foundation, Ford Foundation, Mexican Government, International Centre for Wheat and maize Improvement [CIMMYT] and International Rice Research Institute, Philippines.	It is funded by private seed companies. Six seed companies *viz.*., Monsanto, Syngenta, Bayer Crop Science, DuPont, Dow and BASF Plant Science control practically the entire research and output in the field of transgenic plants.
4	Accessibility of the technology	It is accessible to all the farmers.	The seed belongs to the company, which strictly controls its movement.
5	Application of Intellectual Property Rights	Intellectual Property Rights are not applicable *i.e.* patenting is not done	Intellectual Property Rights are applicable *i.e.* patenting is done.
6	Gene Transfer	The gene transfer was within the species.	The gene transfer is possible between two totally unrelated organisms or across any species.

Sl.No.	Particulars	Green Revolution	Gene Revolution
7	Methods Involved	Conventional methods such as selection, hybridization and mutation.	Highly sophisticated biotechnological approaches are used for development of transgenic plants
8	Transparency	There is 100 per cent transparency in the green revolution	The information is kept secret by the seed companies responsible for developing the technology.
9	Benefits	The green revolution has brought lot of benefit to the farmers nation and consumers.	The gene revolution is expected to bring even greater benefits to the farmers and the poor.
10	Bio-safety measures	Not required	Required
11	Bio-ethical measures	Not required	Required
12	Organic farming of end product	Permitted	Not permitted
13	End product	Non transgenic variety	Transgenic cultivar
14	Evolution of super weeds	Not possible	Possible
15	Technical Skill required	Moderate	Very high
16	Initial cost of Infra structure	Low	Very high
17	Sexual process for gene transfer	Involved	Not involved
18	Laboratory required	Ordinary	Sophisticated
19	Time required to develop new variety	10-12 years	4-5 years
20	Trans gene	Not involved	Involved
21	Characters improved	Mainly productivity and plant stature.	Mainly insect resistance and herbicide resistance.
22.	Crops covered	Mainly cereals, *viz*.. wheat and rice.	All major field crops including vegetable crops.

Advantages of GM Technology

The cultivation of GM crops has played significant role in improving yield, in some cases food quality and reducing use of pesticides or particularly harmful herbicides. It is expected that GM crops can revolutionize world agriculture, particularly in developing countries. This technology will substantially improve food security, reduce malnutrition, and increase rural income, and in some cases even reduce environmental pollutants.

Five criteria are considered important in evaluation of a technology for agricultural revolution. These criteria include (i) benefit to farmers, (ii) increase in production, (iii) adoption by farmers and acceptance by the consumers, (iv) coordination among technology providers, policy makers and users, and (v) the technology should be affordable by users.

1. The new technology should provide a net benefit to farmers.

2. The technology should substantially improve agricultural production, food nutrition, or both; or substantially decrease necessary inputs such as fertilizer, insecticides, herbicides and water.
3. People should be willing to adopt the new technologies, and consumers should accept the products of the such technology.
4. There should be cooperation among technology provider, policy makers and users.
5. The technology should be affordable by the users and sustainable without public subsidization.

The GM crops meet all five criteria for an agricultural revolution on regional scale. In the United States, Canada, China, and Argentina, for example, genetically modified varieties of soybeans, corn, and cotton now constitute about 30 to 80 percent of total plantings of these crops and are beneficial to the growers. In other nations, such as India and South Africa, farmers have recently begun to plant GM crops and have made the beginnings of realizing benefits of Gene Revolution. However, the revolution has yet to occur on a global scale.

The Green Revolution that had its origins in the 1940s, and reached its peak in the 1970s, continues to affect agricultural practices today. Initially, the objective of the Green Revolution was to increase food production in regions of the world facing impending massive malnutrition. In the post-World War II era, scientists and policy makers considered those regions to be Latin America and Asia. The introduction of high-yield varieties (HYVs) of crop seed, along with pesticides, fertilizers, and irrigation systems, transformed agriculture in these two continents.

The Future of Gene Technology

In future, to achieve agricultural revolution through gene technology, the following points should be given due importance.

1. **Affordable Technology:** The new technology should be affordable by the farmers of developing-world. Moreover, the farmers must understand how to use them.
2. **Research by Public Sector:** There is a need for larger investments in research in the public sector. Partnerships between the public and private sectors can result in more efficient production of GM crops that are useful to the developing world and can expand the accessibility of those crops and their associated technologies to developing-world farmers.
3. **Due Importance to Agriculture:** Agricultural development must be given due importance from a policy perspective in both donor and recipient nations. In view of increasing global population, agricultural development is more necessary than ever to eliminate malnutrition and prevent famine, particularly in sub-Saharan Africa. GM crops seem to be effective means for addressing these problems. The policymakers should work jointly on such issues.

4. **Proper Regulatory System:** Policymakers in the developing world must set regulatory standards that take into consideration the risks as well as the benefits of foods derived from GM crops. This goal is crucial to the cooperation of the many stakeholders that are affected by GM crops and also for the sustainability of the GM crop movement in the near future. Without regulations that explicitly take into account potential benefits to both farmers and consumers, those nations that might stand to benefit most from GM crops may be discouraged from allowing them to be planted. Revised regulations on genetically modified crops must accompany widespread collective policy efforts to revitalize agricultural development.

If the GM crops are free of adverse effects on health and environment, they have the potential to provide benefits to farmers and consumers around the globe. Such crops will reduce the use of potentially harmful chemicals or scarce water supplies for agriculture. It can then indeed become a true "Gene Revolution."

The effects of GM crops on health and environment are uncertain. The cultivation of such crops can have unintended adverse effects on both animal health and the environment.

SUMMARY

Gene revolution refers to quantum jump in the productivity of various field crops through use of transgenic cultivars. It has been achieved through the application of Agricultural Biotechnology. Hence agricultural biotechnology is referred to as the 'Evergreen Revolution' or the 'Gene Revolution. Modern biotechnology plays a vital role in the Gene Revolution.

The main features of gene revolution or ever green revolution in terms of ownership, funding source, accessibility of the technology, application of IPR, gene transfer, method employed, transparency, benefits, bio-safety, bioethical measures, *etc.* in relation to green revolution have been discussed.

There are some similarities and some resemblances between gene revolution and green revolution. These have been presented in a tabular form.

Five criteria are considered important in evaluation of a technology for agricultural revolution. These criteria include (i) benefit to farmers, (ii) increase in production, (iii) adoption by farmers and acceptance by the consumers, (iv) coordination among technology provider, policy makers and users, and (v) the technology should be affordable by users. The GM crops meet all five criteria for an agricultural revolution on regional scale.

To achieve agricultural revolution through gene technology, the following points should be given due importance. The technology should be affordable, the research should be taken up by public sector, due importance should be given to agriculture and proper regulatory system.

The effects of GM crops on health and environment are uncertain. The cultivation of such crops can have unintended adverse effects on both animal health and the environment.

QUESTIONS

1. **Define gene revolution and describe briefly its important features.**
2. **Give in brief comparison between green revolution and gene revolution in tabular form.**
3. **Explain briefly advantages of gene technology.**
4. **Discuss in brief future prospects of gene technology in relation to crop improvement.**
5. **Write short notes on the following:**

   (i) Gene revolution (ii) Green revolution
6. **Discuss gene revolution on the basis of following:**

   (i) Funding Agency (ii) Ownership of technology

   (iii) Transparency (iv) Benefit sharing

Chapter 18

Nanotechnology in Crop Improvement

INTRODUCTION

The term nanotechnology was defined in 1974 by a professor from Tokyo Science University, Norio Taniguchi. By definition nanotechnology deals with units that are smaller than 100 nanometers. Nanotechnology has been defined in various ways. Some definitions of nanotechnology are as follows:

(i) Nanotechnology, in short called nanotech, is the branch of engineering that deals with things smaller than 100 nanometers (especially with the manipulation of individual molecules)

(ii) Nanotechnology, sometimes shortened to nanotech, refers to a field of applied science whose theme is the control of matter on an atomic and molecular scale. Generally nanotechnology deals with structures 100 nanometers or smaller, and involves developing materials or devices within that size.

(iii) It is the science and technology of creating nanoparticles and of manufacturing machines which have sizes within the range of 0.1to100 nanometer

Main Features

Nanotechnology deals with the creation and use of objects at the Nano scale, up to 100 nanometers in size. One Nano-meter is one-millionth of a millimeter or one billionth of a meter or one thousand-millionth of a meter. The main points related to nanotechnology are listed below:

(i) Nanotechnology creates and uses structures, devices, and systems that have novel properties.

(ii) It is the manipulation or manufacture of material at the molecular level, or one billionth of a meter.

(iii) It includes the ability to devise self-replicating machines, robots, and computers that are molecular sized,

(iv) It also includes Nano-delivery systems for drugs, and quantum and molecular computing.

(v) It is the science and technology of building electronic circuits and devices from single atoms and molecules.

(vi) Nanotechnology involves working with particles at nanometer scale [at the atomic or molecular level].

(vii) It makes it possible to arrange atoms and molecules in certain defined structures.

(viii) It deals with manipulating and exploiting the properties of matter at a molecular level.

(ix) It is the science of particles that are measured by a nanometer, which is one billionth of a meter.

Brief History

Idea of the use of nanotechnology was first given in 1959 by a well-known physicist Richard Feynman in his speech at an American Physical Society meeting at California Institute of Technology (Caltech) on December 29, 1959. His speech was completely theoretical and seemingly fantastic because at that it was not possible to manipulate single atoms or molecules as they were far too small for manipulation.

He described that the laws of physics do not limit our ability to manipulate single atoms and molecules. However, it was our lack of the appropriate methods for doing so. He correctly predicted that the time would come in which atomically precise manipulation of matter would inevitably arrive.

Applications

Nanotechnology has a number of potential applications in diversity of fields such as (i) medicine, (ii) filtration, (iii) energy, (iv) displays, (v) consumer goods, (vi) agriculture, and (vii) crop improvement. First applications of nanotechnology have been discussed in crop improvement and agriculture and then applications in other fields are presented in Table 18-1.

TABLE 18-1: Applications of Nanotechnology

Sl.No.	*Main Field*	*Applications*
1	Agricultural Biotechnology [Crop Improvement]	(i) Gene delivery (ii) Fine contro l(iii) Multiple gene delivery (iv) Study of gene function

Sl.No.	Main Field	Applications
2	Agriculture	(i) Detection of disease (ii) Treatment of disease (iii) Better nutrient absorption (iv) Better use of insecticides and herbicides (v) Protection of environment (vi) Precision farming
3	Medicine	(i) Drug delivery (ii) Tissue engineering:
4	Filtration	Qaste-water treatment, air purification
5	Energy	(i) Increasing the efficiency of energy production (ii) Recycling of batteries
6	Displays	Carbon nanotubes
7	Consumer goods	Foods, house hold, optics, textiles and cosmetics.

Applications in Crop Improvement

Nanotechnology has practical applications in crop improvement through its use in agricultural biotechnology. A team of Iowa State University plant scientists and chemists [Brian Trewyn, Francois Torney, Kan Wang and Victor Lin] first used nanotechnology to penetrate rigid plant cell walls and deliver DNA and chemicals with precise control. They successfully used nanotechnology to penetrate plant cell walls and deliver a gene and a chemical that triggers its expression. This application of nanotechnology to plant biology and agricultural biotechnology has resulted in a significant breakthrough in delivering specific gene into plant cells. The main applications of nanotechnology in agricultural biotechnology include (i) gene delivery, (ii) controlled gene delivery, (iii) multiple gene delivery, and (iv)study the gene function. These are briefly discussed below:

1. **Gene Delivery**: It permits successful delivery of a gene into a plant cell. Till recently, we were at nature's mercy when we delivered a gene into a cell," Lin said. "There has been no precise control as to whether the cells will actually incorporate the gene and express the consequent protein. With this technology, we may be able to control the whole sequence in the future." And once the gene is inside the plant cell wall, it opens up "whole new possibilities. In the future, scientists could use the new technology to deliver imaging agents or chemicals inside cell walls. This would provide plant biologists with a window into intracellular events.
2. **Controlled Technique**: We can bring in a gene and induce it in a controlled manner at the same time and at the same location. That has never been done before" [Wang}.
3. **Multiple Gene Delivery:** The most tremendous advantage is that several genes can be delivered into a plant cell at the same time and released them

whenever required [Torney said]. With the mesoporous nanoparticles, it is possible to deliver two biogenic species at the same time.

4. **Study of Gene Function:** The controlled release of gene in to plant cell will improve the ability to study gene function in plants. In a separate process, chemicals are used to activate the gene's function. The process is imprecise and the chemicals could be toxic to the plant.

In plant, transformation mostly occurs with the use of a gene gun. In order to use the gene gun to introduce the nanoparticles to walled plant cells, the chemists made another useful modification on the particle surface. They synthesized even smaller gold particles to cap the nanoparticles. These "golden gates" not only prevented chemical leakage, but also added weight to the nanoparticles, enabling their delivery into the plant cell with the standard gene gun.

Achievements

The biologists have successfully used the nanotechnology to introduce DNA and chemicals to Arabidopsis, tobacco and corn plants.

"The team also found a chemical that made the nanoparticle look yummy to the plant cells so they would swallow the particles," The nanoparticles were swallowed by the plant protoplasts [naked cells].

Application in Agriculture

There are new challenges in agricultural sector. Such challenges include a growing demand for healthy and safe food, an increasing risk of disease, and threats to agricultural and fishery production from changing weather patterns. Nanotechnology has the potential to revolutionize the agricultural and food industry with new tools. Important applications of nanotechnology in the field of agriculture include (i) treatment of diseases, (ii) detection of diseases, (iii) better nutrient absorption, (iv)better use of insecticides and herbicides, and (v) protection of environment. All these have been discussed below:

1. **Treatment of Diseases:** It provides molecular treatment of diseases.
2. **Detection of Diseases:** It helps in rapid detection of diseases. Smart sensors and smart delivery systems will help the agricultural industry combat viruses and other crop pathogens.
3. **Better Nutrient Absorption:** It enhances the ability of plants to absorb nutrients *etc.*
4. **Better Use of Insecticides and Herbicides:** In the near future nano structured catalysts will be available which will increase the efficiency of pesticides and herbicides, allowing lower doses to be used.
5. **Protection of Environment:** Nanotechnology will also protect the environment indirectly through the use of alternative (renewable) energy supplies, and filters or catalysts to reduce pollution and clean-up existing pollutants.

An agricultural methodology widely used in the USA, Europe and Japan, which efficiently utilized modern technology for crop management, is called Controlled Environment Agriculture (CEA). CEA is an advanced and intensive form of hydroponically-based agriculture. Plants are grown within a controlled environment so that horticultural practices can be optimized. The computerized system monitors and regulates localized environments such as fields of crops. CEA technology, as it exists today, provides an excellent platform for the introduction of nanotechnology to agriculture. In CEA, nano-technological devices can tremendously improve the grower's ability to determine the best time of harvest for the crop, the vitality of the crop, and food security issues, such as microbial or chemical contamination.

6. **Precision Farming:** There are two main applications of nanotechnology in the precision farming, *viz*.. (i) reduction in the cost of production, and (ii) reduction in pollution. These are discussed below:

   (i) **Reduction in Cost of Production:** The major goal of precision farming is to maximize output (*i.e.* crop yields) and minimize input (*i.e.* fertilizers, pesticides, herbicides, *etc.*) by monitoring environmental variables and applying targeted action. Precision farming makes use of computers, global satellite positioning systems, and remote sensing devices to measure highly localized environmental conditions. Thus it determines whether crops are growing at maximum efficiency or precisely identifies the nature and location of problems. By using centralized data to determine soil conditions and plant development, seeding, fertilizer, chemical and water use can be fine-tuned to lower production costs and potentially increase production- all benefiting the farmer.

   (ii) **Reduction in Pollution:** Precision farming can also help to reduce agricultural waste and thus keep environmental pollution to a minimum. Although not fully implemented yet, tiny sensors and monitoring systems enabled by nanotechnology will have a large impact on future precision farming methodologies. One of the major roles for nanotechnology-enabled devices will be the increased use of autonomous sensors linked into a GPS system for real-time monitoring. These Nano-sensors could be distributed throughout the field where they can monitor soil conditions and crop growth. Wireless sensors are already being used in certain parts of the USA and Australia. For example, one of the Californian vineyards, Pickberry, in Sonoma County has installed wifi systems with the help of the IT Company, Accenture. The initial cost of setting up such a system is justified by the fact that it enables the best grapes to be grown which in turn produce finer wines, which command a premium price. The use of such wireless networks is of course not restricted to vineyards only. This technique can also be used in other crops.

Applications in Other Fields

Besides above, nanotechnology has wide practical applications in a diversity of fields such as (i) medicine, (ii) filtration, (iii) energy, (iv) displays,and (v) consumer goods. These are briefly presented below:

1. Medicine

(i) **Drug delivery:** The overall drug consumption and side-effects can be lowered significantly by depositing the active agent in the morbid region only and in no higher dose than needed. This highly selective approach reduces costs and human suffering. An example can be found in dendrimers and Nano-porous materials. They could hold small drug molecules and transporting them to the desired location.

(ii) **Tissue engineering:** Nanotechnology can help to reproduce or to repair damaged tissue. Tissue engineering might replace today's conventional treatments, *e.g.* transplantation of organs or artificial implants.

2. Filtration

A strong influence of Nano-chemistry on waste-water treatment, air purification and energy storage devices is to be expected. Nano-filtration is mainly used for the removal of ions or the separation of different fluids.

Magnetic Nano-particles offer an effective and reliable method to remove heavy metal contaminants from waste water by making use of magnetic separation techniques. Using Nano-scale particles increases the efficiency to absorb the contaminants and is comparatively inexpensive compared to traditional precipitation and filtration methods.

3. Energy

(i) **Increasing the efficiency of energy production:** Nanotechnology could improve combustion by designing specific catalysts with maximized surface area. Scientists have recently developed tetrad-shaped Nano-particles that, when applied to a surface, instantly transform it into a solar collector.

(ii) **Recycling of batteries:** The use of batteries with higher energy content or the use of rechargeable batteries or super-capacitors with higher rate of recharging using Nano-materials could be helpful for the battery disposal problem.

4. Displays

The production of displays with low energy consumption could be accomplished using carbon nanotubes (CNT). Carbon nanotubes can be electrically conductive and due to their small diameter of several nanometers, they can be used as field emitters with extremely high efficiency.

5. Consumer Goods

In case of consumer goods, nanotechnology has wide applications in foods, house hold, optics, textiles and cosmetics as discussed below:

(i) **Foods:** Nanotechnology can be applied in the production, processing, safety and packaging of food. A Nano-composite coating process could improve food packaging by placing anti-microbial agents directly on the surface of the coated film. Nano-composites could increase or decrease gas permeability of different fillers as is needed for different products.

They can also improve the mechanical and heat-resistance properties and lower the oxygen transmission rate. Research is being performed to apply nanotechnology to the detection of chemical and biological substances for sensing biochemical changes in foods.

(ii) **Household:** The most prominent application of nanotechnology in the household is self-cleaning or easy-to-clean surfaces on ceramics or glasses. Nano-ceramic particles have improved the smoothness and heat resistance of common household equipment.

(iii) **Optics:** The first sunglasses using protective and antireflective ultrathin polymer coatings are on the market. For optics, nanotechnology also offers scratch resistant surface coatings based on Nano-composites. Nano-optics could allow for an increase in precision of pupil repair and other types of laser eye surgery.

(iv) **Textiles:** The use of engineered Nano-fibers already makes clothes water- and stain-repellent or wrinkle-free. Textiles with a Nano-technological finish can be washed less frequently and at lower temperatures. Nanotechnology has been used to integrate tiny carbon particles membrane and guarantee full-surface protection from electrostatic charges for the wearer. A military application could be in camouflage where Nano-cameras mixed with Nano-displays could create an invisibility coat, acting like the skin of a Chameleon.

(v) **Cosmetics:** One field of application is in sunscreens. The traditional chemical UV protection approach suffers from its poor long-term stability. A sunscreen based on mineral nanoparticles such as titanium dioxide offer several advantages. Titanium oxide nanoparticles have a comparable UV protection property as the bulk material, but lose the cosmetically undesirable whitening as the particle size is decreased.

Implications of Nanotechnology

Nanotechnology has a number of potential useful applications, but the side effect of this advance technology can not be ignored. The effects of Nano-materials on human health and environment are briefly presented below:

1. **Effects on Human Health:** There are scientific evidences which demonstrate that some Nano-materials are toxic to humans or the environment. The smaller a particle, the greater is its surface area to volume ratio and the

higher is its chemical reactivity and biological activity. Nano-materials toxicity may result in oxidative stress, inflammation, and consequent damage to proteins, membranes and DNA.

The extremely small size of Nano-materials also means that they are much more readily taken up by the human body than larger sized particles. Size is therefore a key factor in determining the potential toxicity of a particle. However it is not the only important factor. Other properties of Nano-materials that influence toxicity include: chemical composition, shape, surface structure, surface charge, aggregation and solubility, and the presence or absence of functional groups of other chemicals.

The health risks of Nano-materials are of particular concern for workers who may face occupational exposure to Nano-materials at higher levels, and on a more routine basis, than the general public.

2. **Effects on Environment:** The disposal of Nano-particles will pollute the environment and affect the ecosystem. Hence, the authorities should take appropriate action to minimize the side effects of Nano-materials. There should be awareness in the public about the side effects of nanotechnology.

SUMMARY

The term nanotechnology was defined in 1974 by a professor from Tokyo Science University, Norio Taniguchi. By definition nanotechnology deals with units that are smaller than 100 nanometers. Nanotechnology has been defined in various ways. Some definitions of nanotechnology are as follows:

It is the branch of engineering that deals with things smaller than 100 nanometers (especially with the manipulation of individual molecules)

Nanotechnology deals with the creation and use of objects at the Nano scale, up to 100 nanometers in size. One Nano-meter is one-millionth of a millimeter or one billionth of a meter or one thousand-millionth of a meter. The main points related to nanotechnology have been discussed.

Nanotechnology has a number of potential applications in diversity of fields such as (i) medicine, (ii) filtration, (iii) energy, (iv) displays, (v) consumer goods,(vi) agriculture, and (vii) crop improvement. All these have been discussed.

Important applications of nanotechnology in the field of agriculture include (i) treatment of diseases, (ii) detection of diseases, (iii) better nutrient absorption, (iv) better use of insecticides and herbicides, and (v) protection of environment. All these have been discussed.

There are two main applications of nanotechnology in the precision farming, *viz..* (i) reduction in the cost of production, and (ii) reduction in pollution. These have been discussed.

The main applications of nanotechnology in agricultural biotechnology include (i) gene delivery, (ii) controlled gene delivery, (iii) multiple gene delivery, and (iv) study of gene function. All these have been discussed.

Applications in the field of (i) medicine, (ii) filtration, (iii) energy, (iv) displays, and (v) consumer goods have been briefly presented.

Nanotechnology has a number of potential useful applications, but the side effects of this advance technology can not be ignored. The effects of Nano-materials on human health and environment have been discussed.

QUESTIONS

1. **Define Nanotechnology and describe briefly various applications of nanotechnology.**
2. **What are the important applications of Nanotechnology in crop improvement?**
3. **Explain briefly applications of nanotechnology in agriculture.**
4. **Discuss briefly general applications of nanotechnology.**
5. **Explain applications of nanotechnology in the field of:**

   (i) Medicine (ii) Agriculture

   (iii) Crop Improvement (iv) Textiles

7. **What are side effects of nanotechnology?**
8. **Explain nanotechnology in relation to precision farming.**
9. **Write short notes on the following:**

   (i) Nanotechnology (ii) Nano-particles

   (iii) Nanometer (iv) Nano-engineering

Chapter 19

Reverse Breeding [RB]

INTRODUCTION

Reverse breeding can be defined as a biotech based method of obtaining homozygous parental lines from heterozygous genotype for use in conventional (classical) plant breeding. The main objective of reverse breeding is to generate homozygous parental lines that can be crossed to recreate a desired heterozygous genotype. The concept of reverse breeding was developed by Dirks *et al.* (2003) working with *Arabidopsis thaliana* (a close relative of mustard).

Production of parental lines from a heterozygous plant is one of the important goals in plant breeding. Reverse breeding (RB) is a novel plant breeding technique to achieve this goal. RB generates perfectly homozygous parental lines through engineered meiosis. This technique is based on reducing genetic recombination in the selected heterozygote by eliminating meiotic crossing over. Male or female spores obtained from such plants contain combinations of non-recombinant parental chromosomes which can be cultured *in vitro* to generate homozygous doubled haploid plants (DHs). From these DHs, complementary parents can be selected and used in perpetual reconstitution of heterozygous genotype.

There are two methods that allow the fixation of elite germplasm. These are apomixis and reverse breeding. Apomixis helps in fixation of elite germplasm lines but does not permit further improvement of such lines. Moreover, it is difficult to obtain apomictic line at will in a plant species. Reverse breeding is useful in fixation of elite germplasm lines as well as in further improvement of such lines. Thus, RB is expected to play key role in future plant breeding.

Important Features

There are several important features of reverse plant breeding. Some main points related to reverse breeding are listed as follows:

1. **Use of Cytogenetic Techniques**: Reverse breeding involves cytogenetic techniques such as fluorescent *in situ* hybridization. In reverse breeding both male as well as female meiosis are controlled to get gametes with parental chromosomes.
2. **Use of Molecular Markers:** It also makes use of molecular markers for identification of homozygous parental lines. SNP markers are commonly used for this purpose.
3. **Non Recombinant Meiosis:** It involves achiasmatic meiosis or meiosis free from recombination or crossing over. In other words, reverse breeding is based on transmission of parental chromosomes with intact genes.
4. **Gene Linkage:** Reverse breeding is based on the absolute linkage of genes present on the homologous parental chromosomes. Reverse breeding relies on the effective suppression of meiotic crossovers. Thus, genes that suppress crossing-over are useful.
5. **Creation of Chromosome Substitution Lines:** With reverse breeding, one can obtain all possible chromosome-substitution lines in a short period of time possibly in two generations.
6. **Use of RNA Interference:** The knockdown of gene expression, essential for RB, can be achieved by targeting genes using RNA interference (RNAi).

Steps Involved

The reverse breeding technique involves six important steps *viz..*, selection of heterozygous genotype, inhibition of crossing over, production of balanced gametes, selection of balanced gametes, conversion of haploid gametes in to diploid adults and use of such genotypes in classical plant breeding. These steps are discussed as follows:

1. **Selection of Heterozygous Genotype:** The first important step in reverse breeding is selection of a fully heterozygous F_1 resulting from a cross between two homozygous parents. Thus RB starts with a heterozygote in which meiotic recombination can be suppressed. Reverse breeding can be used as advanced breeding tool.
2. **Inhibition of Recombination:** Reverse breeding, as stated above, involves cytological techniques. Such techniques help in linking the parental chromosomes. The crossing can be inhibited by various techniques of genetic engineering.
3. **Production of Balanced Gametes:** The third important step is production of balanced gametes with parental chromosomes. Balanced gametes can be produced in the absence of chiasmata. Gametes are of two types, *viz..*, unbalanced gametes and balanced gametes. The gametes having aneuploidy are known as unbalanced gametes. On the other hands, gametes with normal chromosomes are referred to as balanced gametes. In the absence of crossovers, non-recombinant chromosomes

segregate randomly. This segregation is generally unbalanced, leading to aneuploidy in gametes. This may result in production of sterile phenotype of achiasmatic plants. However, by chance balanced gametes are also produced at a very low frequency. In *Arabidopsis thaliana* (n = 5) frequency of balanced gametes is 3.25 per cent. This predicts ample scope for production of viable pollen in the absence of crossovers, even for species with higher chromosome numbers.

4. **Selection of balanced Gametes:** The balanced parental gametes can be identified by cytological techniques such as fluorescent *in situ* hybridization. These gametes are used for production of diploid adult plants.
5. **Conversion of Haploid gametes into Diploid Adults:** The fifth important step in reverse breeding is to convert haploid gametes, carrying non-recombined chromosomes, into homozygous diploid adults. This can be achieved through different methods depending on the plant species. In *A. thaliana*, haploid plants can be produced by centromere-mediated genome elimination. These haploids can be converted into fertile diploids (doubled haploids) by collecting the rare seeds from self-pollination of the haploids or through colchicine treatment of haploid plants.
6. **Use in Classical Breeding:** The diploid plants developed from gametes containing parental chromosomes are used in classical plant breeding for developing hybrids as well as homozygous cultivars for commercial cultivation.

Forward and Reverse Breeding

Forward and reverse breeding are apposite approaches of each other. Forward breeding is widely used for crop improvement whereas reverse breeding is of recent origin and is less popular. Important features of both these approaches are presented as follows:

Forward Breeding

The area of breeding which promotes genetic recombination and reduces recovery of parental type is called forward breeding. It includes classical breeding. Thus there some important features of forward breeding which are listed as follows:

1. Genetic recombination is enhanced.
2. Recovery of parental genotypes is reduced.
3. It includes conventional breeding techniques.
4. It requires simple laboratory.
5. It involves moderate technical skill.
6. The initial cost involved is low.
7. It requires more time.

Reverse Breeding

The area of breeding which suppresses genetic recombination and promotes recovery of parental types is called reverse breeding. It includes molecular breeding. Thus there some important features of reverse breeding, which are listed as follows:

1. Genetic recombination is suppressed.
2. Recovery of parental genotypes is promoted.
3. It includes molecular techniques.
4. It requires sophisticated laboratory.
5. It involves high technical skill.
6. The initial cost involved is high.
7. It requires less time.

TABLE 19-1: Comparison of Forward and Reverse Breeding

Sl.No.	*Particulars*	*Forward Breeding*	*Reverse Breeding*
1	Promotes	Recombination	Parental genotypes
2	Reduces	Parental types	Recombinants
3	Technique used	Conventional	Molecular
4	Skill required	Moderate	High
5	Laboratory required	Simple	Sophisticated
6	Cost involved	Low	High
7	Time required	More	Less

Practical Applications

There are six important applications of reverse breeding, *viz.*., recovery of homozygous parental lines, perpetuation of heterozygous lines, improvement of parental lines, and creation of chromosome substitution lines. These are briefly discussed as follows:

1. **Recovery of Homozygous Parental Lines:** Reverse breeding allows recovery of homozygous parental lines from any heterozygous genotype. Such selected homozygous parental types can be used for reconstitution of heterozygous populations (hybrids) or homozygous lines (cultivars). One important application is the production of complementary homozygous lines that can be used to generate specific F1 hybrids. Reverse breeding can be used to fix unknown heterozygotes.
2. **Perpetuation of Heterozygous lines:** Reverse breeding permits continuous production of heterozygous genotype as F_1 which is not possible through heterosis breeding. This overcomes the limitation of heterosis breeding in which hybrids are generated by controlled crossing that too at a very low frequency. The genomes of uncharacterized heterozygous plants can now be fixed in complementing immortal lines. This ability to fix heterozygous

genomes resembles apomixis, in which heterozygotes reproduce clonally through seeds, but is fundamentally different. However, apomictic lines not be improved further by conventional hybrid breeding techniques.

3. **Improvement of Parental Lines:** Reverse breeding generates homozygous parental lines for the selected heterozygote. Thus it allows the improvement of the heterozygote through improvement of the individual parental lines by traditional breeding methods such as backcrossing, mutagenesis and so on.

4. **Creation of Chromosome-substitution Lines:** When RB is applied to F_1 heterozygotes, it is possible to generate chromosome substitution lines. Thus, reverse breeding permits creation of chromosome substitution lines. Chromosome-substitution lines are valuable tools in many breeding applications, such as trait mapping, the study of epistatic interactions and targeted inbreeding. Chromosome-substitution lines in *Arabidopsis thaliana* have been generated through traditional crossing. However, this method requires more generations of crossing and extensive genotyping to WT meiosis.

   In another application, RB can be applied to plants of known background. If crossing over is eliminated in the F_1 hybrid rather than the F_2 generation, RB can be used to generate chromosome substitution lines. These lines contain one or more chromosomes from one parent in the background of the other parent. By backcrossing the chromosome substitution lines to the original parental lines, one can obtain populations that segregate only for the heterozygous chromosome(s). **Reverse breeding**, in theory, allows the re-shuffling of chromosomes between two homozygous plants in all possible ways.

5. **Use with CMS Lines:** RB is fully compatible with commercial CMS lines that are frequently used in modern agriculture. In several vegetable crops such as cabbages and carrots, breeders make use of cytoplasmic male sterility (CMS) (Chase, 2007). In these systems, the presence of male sterility presents a special challenge to RB. In these cases, gyno-genesis rather than andro-genesis can be used to obtain DH plants. This is perfectly compatible with RB in the sense that the chromosomes from the maintainer line can be recovered directly in the cytoplasm of the sterile line in one step. Gynogenesis has been described in several crops such as *Brassica*, maize, sugar beet, cucumber, melon, rice, onion, sunflower, and barley (Keller and Korzun, 1996). However, this technique was given up when anther and microspore culture techniques were developed. In cases where the efficiency of gynogenesis is too low, it is possible to cross the male sterile (A) lines with maintainer lines (B) that carry one copy of a restorer gene. The AB combination will be fertile and RB can be performed. In rice, restorer genes have been successfully transformed (Wang *et al.*, 2006). It should therefore be possible to use a restorer gene and a gene for crossover suppression in the same vector (both transgenes) and perform RB in a 'double suppressed' (CMS and crossover) background.

6. **Reconstruction of Heterozygous Germplasm:** For crops where an extensive collection of breeding lines is still lacking, RB can accelerate the development of varieties. In these crops, superior heterozygous plants can be propagated without prior knowledge of their genetic constitution.

Reverse Breeding and Marker Assisted Breeding

Especially in combination with (high throughput-) genotyping, reverse breeding becomes a versatile tool. Evidently, high throughput genotyping speeds up the process of identification of complementing parents in populations of DHs in early stages. But perhaps more powerful is its use in the study of gene interactions of the various heterozygous inbred families (HIFs) that can be produced by crossing and backcrossing the products of RB. The screening of populations that segregate for traits on a single chromosome allow the quick identification of QTLs, when genotyping is combined with –for example- transcriptome or metabolome profiling. Such HIFs further aid the generation of chromosome specific linkage maps and the fine mapping of genes and alleles. RB can as such provide highly valuable insights into the nature of heterotic effects.

Limitations

There are some limitations which affect the application or usefulness of reverse breeding. These are segregation distortion, low frequency, effect of chromosome number and polyploidy and doubled haploids. These are briefly discussed as follows:

1. **Segregation Distortion:** In reverse breeding, linkage is absolute for all loci on a chromosome. However, any distorting locus would affect transmission of the whole chromosome. Segregation distortion is observed during meiosis which also poses problems in the selection of balanced gametes with parental chromosome.
2. **Low Frequency:** The frequency of desirable gametes is extremely low. This limits the application of this technique in practical plant breeding.
3. **Effect of Chromosome Number:** Reverse breeding might be difficult to adopt in crops with higher chromosome numbers (for example, soybean) as well as in polyploids such as canola, cotton, wheat and potato. The expected low production of balanced gametes in crops with higher chromosome numbers is a possible bottleneck in reverse breeding. However, successful reverse breeding does not require the complete knockdown of crossovers, and the occurrence of a few crossovers might be beneficial, as non-recombinant chromatids are still present in bivalent pairs with a single crossover. In rice ($n = 12$), for example, allowing three crossovers will increase the chance of finding a balanced gamete.
4. **Effect of Polyploidy:** The technique however is limited to crops with a haploid chromosome number of 12 or less and in which spores can be regenerated into DHs. In polyploids or species with high chromosome numbers, another reconstruction method has been proposed that is based

on the omission of the second meiotic division, leading to unreduced second division restitution (SDR) spores. The use of these SDR spores enables the near reconstruction of desired phenotypes, and also provides the possibility of obtaining chromosome substitution lines (Van Dun and Dirks, 2006).

5. **Application:** Use of reverse breeding is limited to those crops where DH technology is common practice. For the great majority of crop species this technology is well established and professional breeding companies routinely use such techniques in their breeding programs.

In the complete absence of meiotic recombination one polymorphic molecular marker per chromosome would suffice to genotype every DH since the entire chromosome would behave as a single linkage block. In the presence of any residual crossovers, two markers (as distally located as possible) are required per chromosome.

Doubled Haploids

Doubled haploid plants resulting from achiasmatic meiosis can be obtained from unfertilized ovules (gynogenesis) or from microspore and anther cultures (androgenesis), according to well-established protocols that have been developed for a variety of plant species including crops (Jain *et al.*, 1996). The efficiency of DH formation from haploid spores is species dependent (Forster *et al.*, 2007). Crossing over is suppressed in this technique and achiasmatic gametes are collected, cultured, and used to generate DHs. The DH lines can then be used to reconstitute the elite heterozygote on a commercial scale.

Chances of Finding Complementing Parents

The maximum number of different DHs obtained from a heterozygous diploid in a RB experiment equals 2^x, with x being the basic chromosome number.

FUTURE SCOPE

The technical feasibility of reverse breeding in *A. thaliana* suggests that it might be possible to apply this technology in crop improvement. Crucially, genes governing meiotic recombination are widely conserved, and haploid generation methods are available for many crops. Although the probability of recovering balanced gametes decreases with increasing chromosome numbers, many agronomically important crops have 12 chromosomes or fewer, and are within the reach of successful reverse breeding. Notable examples include cucumber, onion, broccoli, cauliflower, sugar beet, maize, pea, sorghum, (water-) melon, tomato, pepper, rice, eggplant and so on. Reverse breeding might be difficult to adopt in crops with higher chromosome numbers (for example, soybean) as well as in polyploids such as canola, cotton, wheat and potato.

SUMMARY

Reverse breeding is a biotech based method of obtaining homozygous parental lines from heterozygous genotype for use in conventional (classical) plant breeding.

The main objective of reverse breeding is to generate homozygous parental lines that can be crossed to recreate a desired heterozygous genotype. Important features of reverse breeding have been discussed.

The reverse breeding technique involves six important steps *viz..*, selection of heterozygous genotype, and inhibition of crossing over, production of balanced gametes, selection of balanced gametes and conversion of haploid gametes in to diploid adults and use of such genotypes in classical plant breeding.

There are six important applications of reverse breeding, *viz..*, recovery of homozygous parental lines, perpetuation of heterozygous lines, improvement of parental lines, and creation of chromosome substitution lines. These are briefly discussed.

There are some limitations which affect the application or usefulness of reverse breeding. These are segregation distortion, low frequency, effect of chromosome number and polyploidy and doubled haploids. These have been briefly discussed. Future scope of reverse breeding has been discussed.

QUESTIONS

1. **Describe briefly various important features of reverse breeding.**
2. **Define reverse breeding. How it differs from forward plant breeding?**
3. **What are advantages of reverse breeding over classical plant breeding?**
4. **Describe briefly various steps involved in reverse breeding.**
5. **Discuss briefly various applications of reverse plant breeding.**
6. **Define segregation distortion. Describe briefly various limitations of reverse breeding.**
7. **Describe in brief achievements and future scope of reverse breeding in crop improvement.**
8. **Give a brief comparison of forward breeding and reverse breeding in tabular form.**
9. **Write short notes on the following:**

   (i) Reverse breeding
   (ii) Forward breeding
   (iii) Segregation distortion
   (iv) Doubled Haploids

Chapter 20

Space and Speed Breeding

INTRODUCTION

There are some new emerging plant breeding concepts such as space breeding and speed breeding. A brief account of both these emerging plant breeding concepts is presented as follows:

A. Space Breeding

Space breeding refers to the technique of sending seeds or microorganisms into space (200-400km) on a recoverable satellite or in other recoverable spacecraft. The description of space breeding in terms of brief history, principle involved, purpose, crops used, new varieties developed, characters improved, period of exposure, effect of radiation, advantages, and limitations is presented as follows:

Brief History

The space breeding work was initiated in China in 1980s. The first batch of crop seeds was sent into the orbit through a satellite. This Chinese program continued till 990s in simple way. However, this program got momentum in recent years. Presently, the Chinese government has intensified the space breeding program. This program is carried by the Space Breeding Research Center of the Chinese Academy of Agricultural Sciences. This helped to increase the number of new seed varieties produced by Space breeding from just 20 to 110.

Principe Involved

In the space there are strong-radiation, high-vacuum, low-gravity and alternating magnetic field. Such environment leads to mutagenic changes in the seeds. Some of such mutations may be beneficial. After returning to Earth, mutated seeds are planted in the field and selection is made for plants with high yield and better quality.

Purpose

The main objective of space breeding is to develop a set of new crop cultivars with high yield and good quality, and also creating a number of novel plant genetic resources.

New Varieties Developed

Space breeding carried over 2,000 plant accessions from 133 species. So far, 66 new mutant varieties of crops, including rice, wheat, cotton, rapeseed, sesame, pepper, tomato and alfalfa have been developed by the space-breeding program and officially released in China. Some space bred crop varieties are given in Table 20.1.

TABLE 20-1: Crop Varieties Developed through Space Breeding in China

Sl.No.	*Name of Crop*	*Varieties Developed*
1	Rice	Hangyu1, Huahang1, Eryouhang1, Teyouhang1, Yuehang1, Zhe101, Yuyou1, Teyou175 and Huaxiang7, *etc.* [9]
2	Wheat	Taikong5, Taikong6, Longfumai15, Hangmai96, Longfumai17 and Luyuan301 [6]
3	Cotton	Zhongmiansuo42 and Zhongmiansuo52, [2]
4	Sesame	Zhongzhi11 and Zhongzhi13 [2]
5	Pepper	Yujiao1, Yujiao2, Yujiao3, Yujiao4 and Longjiao9 [5]
6	Tomato	Yufan1 and Yufan2 [2]
7	Alfalfa	Longjing1

China has planted more than 2 million hectares of land with space treated seeds. This resulted in increasing grain yield by over 1 billion kilograms and economic benefit of over 2 billion yuan.

Characters Improved

Space bred varieties have characters of high yield, good quality and multiple resistances. Some useful rare mutants which are expected to make breakthrough in crop yield were also obtained (Qiu *et al.*, 1998; Shi *et al.*2000; Liu *et al.*, 2005a). Thus space breeding has both social and economic benefits.

TABLE 20-2: Characters Improved I different Crops through Space Breeding

Sl.No.	*Name of Crop*	*Character Improved*
1	Rice	4 per cent higher Yield
2	Tomato	Higher protein and sugar content
3	Soybean	11 per cent higher yield
4	Pepper	Bigger fruit size
5	Pumpkin	Giant size
6	Cucumber	Bigger size

The first improved variety of rice variety to be grown from a space seed was Huahang 1, which yielded four per cent more than parent variety. But some of the new varieties of wheat produced by the space breeding program are now being grown commercially in China. One variety of tomato named Taikong 6, developed through space breeding program is more nutritious than normal varieties with more protein content. Other space seed successes, includes a kind of soybean with 11 per cent higher yields than usual, supersized green peppers weighing about 500 gm., gigantic pumpkins, as well as over-sized cucumbers and tomatoes with higher sugar content, whose taste is comparable to oranges.

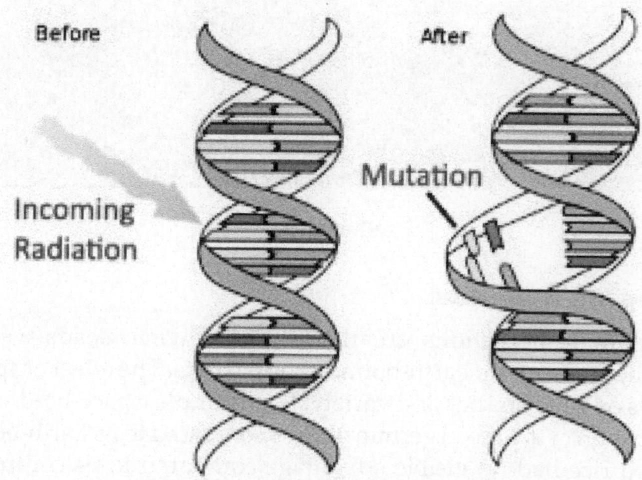

FIGURE 20-1: Radiation Effect on DNA.

Period of Exposure

The period of exposure to space conditions may vary. In China's space breeding experiments, seeds are exposed space conditions for about six days on average. But although the American tomato seeds experienced the same conditions for a vastly longer period of time, experimenters concluded that, fruit yield was unaffected in plants grown from space-exposed seeds. The lucrative space breeding industry has led to a rapid rise in the number of research bases, of which China Governments now has nearly 60 space breeding research bases.

Nuclear Radiation

Chinese researchers are confident that the space breeding program will continue to make breakthroughs and are very happy with the level of government support.

The expenditure on space breeding is much higher than conventional breeding program. It is estimated to be at least $500 USD for each seed lot sent into space. Over 90 percent of seeds are completely unaffected by space flight. Now the breeders are preferring space breeding in combination with traditional seed breeding and without replacing other methods of altering seeds, such as by exposure to nuclear radiation.

Shenzhou

Effects of Space Radiation

Chromosomal aberrations greatly increased when seeds were sent into aerospace compared to their earth-bound counterparts. The effect of space flight on seeds depends on their species and variety. For example, space-bred wheat variety exhibited higher growth in seed germination as compared to its Earth-bound control. But space-bred rice had no visible advantage compared to its control. Positively mutated space bred varieties exhibited better growth and germination than parent material and Earth irradiated counterparts in M1 generation.

Advantages

There are several advantages of space induced mutations than Earth induced mutations. Some advantages are as follows:

1. Space-induced mutation breading can be as a novel and effective way to breed new crop varieties and create novel plant genetic resources.
2. Space breeding has wide mutation spectrum, high frequency of useful genetic variation and short breeding period.
3. Chinese researchers confirm that space travel can change the genetic makeup of plant seeds, producing super-yielding plant varieties when they are eventually planted back on planet earth.
4. Space seeds have produced new kinds of rice, wheat and vegetables which are more productive, larger, and more nutritious than their earthly counterparts.

Limitations

The main limitation of space breeding is that involves huge expenditure. Hence, this technique has not become popular in other countries. In China, due to such restraints, scientists are trying to replicate space condition on Earth to achieve space-born mutations on Earth.

1. One such replication is a magnetic field-free space (MF), which produces an area with a weaker magnetic field than that of Earth. MF treatment produced mutagenic results, and has been used to cultivate new mutant varieties of rice and alfalfa.
2. The other replications of space conditions include irradiation of seeds by a heavy 7 Li-ion beam or mixed high-energy particles.https://en.wikipedia.org/wiki/Mutation_breeding - cite_note-23 These space-bred varieties are already being introduced to the public. In 2011, during the National Lotus Flowers Exhibition in China, a mutant lotus, called the "Outer Space Sun", was shown at the flower show.

B. Speed Breeding

Speed breeding technology is an indoor plant growing technique which shortens harvest time (2-3x faster) of crops like wheat and barley. Speed breeding is also known as accelerated breeding and rapid breeding system. The concept of speed breeding technique was originally developed by NASA in the eighties to help grow crops in space. This can hasten the process of developing new crop varieties. Now many Indian plant scientists are showing interest in exploring the concept of speed breeding. Scientists hope **speed breeding** will accelerate agricultural research and increase food production to meet the demands of a rising population.

This technique uses artificial light and temperature conditions to speed up the crossing and inbreeding of varieties, is receiving increasing attention. The growing human population and a changing environment have raised significant concern for global food security. The current improvement rate of several important crops is inadequate to meet future demand. This slow improvement rate is attributed partly to the long generation times of crop plants.

Main Features

Main features of speed breeding in terms of techniques involved, environment used, approaches used, ways of speed breeding and time required are briefly presented as follows:

1. **Techniques Involved:** Speed breeding' technology involves indoor techniques such as glass house, green house and poly house for growing plants under controlled conditions to grow several generations of selected plants.
2. **Environment Used:** Speed breeding uses an artificial environment with enhanced light duration to create longer daylight regimes to speed up the breeding cycles of photo-insensitive crops. Speed breeding in controlled environment growth chambers can accelerate plant development for research purposes, including phenotyping of adult plant traits, mutant studies and transformation. The use of supplemental lighting using light emitting diodes (LEDs) in a glasshouse environment allows rapid

generation cycling through single seed descent (SSD) and plant density can be scaled-up for large crop improvement programs. Protocols suitable for speed breeding in glasshouses and growth chambers with low cost have been developed. Speed breeding uses enhanced LED lighting and day-long regimes of up to 22 hours to optimise photosynthesis and promote rapid growth of crops. It speeds up the breeding cycle of plants: for example, six generations of wheat can be grown per year, compared to two generations using traditional breeding methods. Speed breeding uses an artificial environment with enhanced light duration to create longer daylight regimes to speed up the breeding cycles of photo-insensitive crops.

It uses supplemental lighting to aid photosynthesis rate in intensive regimes of up to 22 h per day in a glasshouse environment that allows rapid generation cycling through single seed descent and potential for adaptation to larger-scale crop improvement programmes

3. **Approach used:** There is great potential for integrating speed breeding with other modern crop breeding technologies, including high-throughput genotyping, genome editing and genomic selection, accelerating the rate of crop improvement. Speed breeding approach is being carried at ICRISAT with crops like pigeon pea, millets, sorghum and groundnut. These crops are important for food and nutrition security in many countries in Africa and Asia.
4. **Ways of Speed Breeding:** Speed breeding can be carried out in numerous ways, one of which involves extending the duration of plants' daily exposure to light, combined with early seed harvest, to cycle quickly from seed to seed, thereby reducing the generation times for some long-day or day-neutral crops.
5. **Time required:** Speed breeding leads to rapid generation advancement of segregating populations resulting in rapid development of improved cultivars. Conventional breeding takes a longer time for the development of crop varieties with a minimum of 8– 10 years of breeding cycles. The main problem of conventional breeding is the inability to complete more generations in lesser time. A recent report on speed breeding mentions that up to six generations per year could be achieved for bread wheat, durum wheat, barley, pea and chickpea, and four generations for canola

Advantages

A new method for rapid generation advance, called speed breeding, has considerable advantages over DH technology for spring wheat because it provides increased recombination during line development and enables selection in early generations for some traits.

1. **A Rapid Method:** Speed breeding allows researchers to rabidly mobilise the genetic variation found in wild relatives of crops and introduce it into elite varieties that can be grown by farmers. The EU ruling that heavily

regulates gene editing means we are more reliant on speed breeding to grow sturdier, more resilient crops." Speed breeding technology allows six generations of wheat to be grown per year, compared to two generations using traditional breeding methods.

By shortening breeding cycles, the method allows scientists and plant breeders to fast-track genetic improvements such as yield gain, disease resistance and climate resilience in a range of crops such as wheat, barley, oilseed rape and pea.

2. **Rapid Generation Advancement:** Speed breeding' technology shortens the breeding cycle and accelerates crop research through rapid generation advancement. Speed breeding techniques can help plant breeders grow six generations of spring wheat, chickpea and barley and four generations of canola plants in specially modified glasshouses. In comparison, they can grow only one or two generations in the field and two or three in normal glasshouses.

   This technology helps shorten the breeding cycle of new varieties significantly. The speed breeding system is used normally for accelerating the development of plants that have better genetics and then they go to the fields. In fields, they need to be tested for yield, disease resistance and quality,"

   Speed breeding can be used to achieve up to 6 generations per year for spring wheat (*Triticum aestivum*), durum wheat (T. durum), barley (*Hordeum vulgare*), chickpea (*Cicer arietinum*) and pea (*Pisum sativum*), and 4 generations for canola (Brassica napus), instead of 2–3 under normal glasshouse conditions. The technique has also been successfully adapted to oat, various *Brassica* species, grass pea, quinoa, *Medicago truncatula* and *Brachypodium distachyon*.

3. **Early Decision:** Generation time in most plant species represents a bottleneck in applied research programmes and breeding. Tackling this bottleneck means scientists can respond quicker to emerging diseases, changing climate and increased demand for certain traits

4. **Adoption:** Speed breeding technology will become the norm in research institutes: "We know that more and more institutes across the world will be adopting this technology and by sharing these protocols we are providing a pathway for accelerating crop research."

5. **Drought Resistance:** In Australia, speed breeding technology is being used to rapidly cycle genetic improvements to make crops more drought resistant.

6. **Refinement:** This technology needs refinement for use as a research tool. Changes to soil/media composition, lighting, temperature, spacing of plants and premature seed harvest have led to the team cutting down the seed-to-seed generation time in wheat to just eight weeks.

7. **Rapid Gene Transfer:** Development of techniques such as rapid gene discovery and cloning that, alongside speed breeding, would allow crop improvements via a non-GM route.
8. **More Adoption:** In years ahead, crop research and breeding will become more dependent on speed breeding in looking to its various advantages. The ability to work at these scales gives scientists greater opportunities than ever before to breed disease resistant, climate resilient and nutritious crops to feed a growing global population.
9. Speed breeding accelerates the rate at which we can develop hardier, healthier, and more versatile plants in the face of climate change. As climates change, so too does a plant's productivity in a given region. The key to our well-fed future may be a variety of resilient crops that can grow in diverse environmental conditions.
10. Speed breeding will be able to more quickly breed and engineer plants which are more nutritious, resist disease, and which are better adapted to tomorrow's climate."

Limitations

1. The speed-breeding technique has largely been used for research purposes, but is now being adopted by the industries as well. Speed breeding as a platform can be combined with several other technologies such as marker-assisted selection, genomic selection, CRISPR gene editing, *etc.* to get to the end result faster4. With the success of speed breeding particularly in wheat crop, India can also initiate such facilities for quick development of new crop varieties.
2. Speed breeding in the case of photo-insensitive crops like wheat, barley, chickpea and pea greatly reduces the generation time and helps in achieving six generations per year 2, but for the photosensitive crops like soybean, speed breeding is not suitable to speed up the breeding cycles.
3. The enhanced light provided in the speed-breeding facility will not allow the plants to bear flowers in case of photosensitive crops. Therefore, alternative approaches have to be followed to speed up the breeding cycles in such crops.
4. Speed breeding involves greenhouse and poly-house facility, which is used for advancement of F_1 to F_2 and important individual plant selections during off-season (November–February). The few of selected segregating material is also advanced to shorten the varietal development programme.
5. It's in the field that the greenhouse experiments are put to the test. A crop variety may perform well in an artificial environment, where temperature, water supply, and light regimes are easily regulated, but if it doesn't thrive out in the field, where the vast majority of our crops are grown, then it's practically a dud. New wheat varieties must be productive out in the elements for farmers to take them on.

SUMMARY

Space breeding refers to the technique of sending seeds or microorganisms into space (200-400km) on a recoverable satellite or in other recoverable spacecraft. Brief description of space breeding in terms of principle involved, purpose, brief history, crops used, new varieties developed, characters improved, advantages, period of exposure, effects of radiation and limitations is presented. In China, more than 40 improved varieties of different field crops have been released for commercial cultivation through space breeding. Yield and quality have been improved in various crops. Space breeding is a rapid method of crop improvement, but it is very expensive.

Speed breeding technology is an indoor plant growing technique which shortens harvest time (2-3x faster) of crops like wheat and barley. Speed breeding is also known as accelerated breeding and rapid breeding system. The concept of speed breeding technique was originally developed by NASA in the eighties to help grow crops in space. Main features of speed breeding in terms of techniques involved, environment used, approaches used, ways of speed breeding and time required are briefly presented. Advantages and limitations of speed breeding are given.

QUESTIONS

1. **Define space breeding and describe briefly its advantages.**
2. **Discuss in brief crop cultivars developed by space breeding program.**
3. **What is the purpose of space breeding? Explain limitations of such program.**
4. **Define speed breeding and describe briefly its main features.**
5. **Define speed breeding and describe briefly its advantages and limitations.**
6. **Write short notes on the following:**

   (i) Space breeding
   (ii) Recoverable satellite
   (iii) Speed breeding

SUMMARY

QUESTIONS

Appendices

Appendix 1. List of Agricultural Universities in India

Sl.No.	*Name of University*	*Location*	*State*
	I. AGRICULTURAL UNIVERSITIES		
A	**STATE UNIVERSITIES**		
1	Acharya NG Ranga Agricultural University	Hyderabad	Telangana
2	Assam Agricultural University	Jorhat	Assam
3	Bihar Agricultural University	Sabour	Bihar
4	Birsa Agricultural University	Ranchi	Jharkhand
5	Indira Gandhi Krishi Vishwavidyalaya	Raipur	Chhatishgarh
6	Anand Agricultural University	Anand	Gujarat
7	Junagadh Agricultural University	Junagarh	Gujarat
8	Navsari Agricultural University	Navsari	Gujarat
9	Sardarkrushinagar-Dantiwada Agricultural University	Bansakantha	Gujarat
10	Chaudhary Charan Singh Haryana Agricultural University	Hisar	Haryana
11	CSK Himachal Pradesh Krishi Vishvavidyalaya,	Palampur	Himachal Pradesh
12	Sher-E-Kashmir Univ of Agricultural Sciences and Technology	Jammu	Jammu and Kashmir
13	Sher-E-Kashmir Univ of Agricultural Sciences and Technology of Kashmir	Srinagar	Jammu and Kashmir
14	University of Agricultural Sciences	Bangalore	Karnataka
15	University of Agricultural Sciences	Dharwad	Karnataka
16	University of Agricultural Sciences	Raichur	Karnataka
17	Kerala Agricultural University	Vallanikara	Kerala
18	Jawaharlal Nehru Krishi Viswavidyalaya	Jabalpur	Madhya Pradesh
19	Rajmata Vijayaraje Scindia Krishi Viswavidyalaya	Gwalior	Madhya Pradesh
20	Dr Balasaheb Sawant Konkan Krishi Vidyapeeth	Ratnagiri	Maharashtra
21	Dr Panjabrao Deshmukh Krishi Vidyapeeth	Akola	Maharashtra
22	Mahatma Phule Krishi Vidyapeeth	Rahuri	Maharashtra
23	Marathwada Agricultural University	Parbhani	Maharashtra

Sl.No.	*Name of University*	*Location*	*State*
24	Odisha Univ of Agriculture and Technology	Bhubaneswar	Odisha
25	Punjab Agricultural University	Ludhiana	Punjab
26	Maharana Pratap Univ of Agriculture and Technology	Udaipur	Rajasthan
27	Rajasthan Agricultural University	Bikaner	Rajasthan
28	Agriculture University	Kota	Rajasthan
29	Sri Karan Narendra Agriculture University	Jobner	Rajasthan
30	Agriculture University	Jodhpur	Rajasthan
31	Tamil Nadu Agricultural University	Coimbatore	Tamil Nadu
32	Chandra Shekar Azad University of Agriculture and Technology	Kanpur	Uttar Pradesh
33	Acharya Narendra Deva University of Agriculture and Technology	Ayodhaya	Uttar Pradesh
34	Sardar Ballabh Bhai Patel Univ of Agriculture and Technology	Meerut	Uttar Pradesh
35	Manyavar Kashiram University of Agriculture and Technology	Banda	Uttar Pradesh
36	Govind Ballabh Pant University of Agriculture and Technology	Pantnagar	Uttrakhand
37	Bidhan Chandra Krishi Viswavidyalaya	Kalyani	West Bengal
38	Uttar Banga Krishi Viswavidyalaya	Cooch Bihar	West Bengal
B	**CENTRAL UNIVERSITIES**		
1	Central Agricultural University	Imphal	Manipur
2	Rani Laxmibai Central Agricultural University	Jhansi	Uttar Pradesh
3	Rajendra Central Agricultural University	Pusa	Bihar
C	**DEEMED UNIVERSITIES**		
1	Indian Agricultural Research Institute	Pusa	New Delhi
2	National Dairy Research Institute	Karnal	Haryana
3	Indian Veterinary Research Institute	Bareilly	Uttar Pradesh
4	Central Institute for Fisheries Education	Mumbai	Maharashtra
5	Allahabad Agricultural Institute	Prayagraj	Uttar Pradesh

II. ANIMAL SCIENCE UNIVERSITIES

S. N.	Name of University	Location	State
1	Sri Venkateswara Veterinary University	Tirupati	Andhra Pradesh

Sl.No.	Name of University	Location	State
2	Karnataka Veterinary, Animal and Fisheries Sciences University	Bidar	Karnataka
3	Maharashtra Animal Science and Fishery University	Nagpur	Maharashtra
4	Guru Angad Dev Veterinary and Animal Science University	Ludhiana	Punjab
5	Tamil Nadu Veterinary and Animal Science University	Chennai	Tamil Nadu
6	Tamil Nadu Fisheries University	Nagapattinam	Tamil Nadu
7	UP Pandit Deen Dayal Upadhaya Pashu Chikitsa Vigyan Vishwa Vidhyalaya evam Go Anusandhan Sansthan, Mathura	Mathura	Uttar Pradesh
8	West Bengal University of Animal and Fishery Sciences	Kolkata	West Bengal
9	Rajasthan University of Veterinary and Animal Sciences	Bikaner	Rajasthan
10	Kerala University of Fisheries and Ocean Studies	Kochi	Kerala
11	Kerala Veterinary and Animal Sciences University	Wayanand	Kerala
III. HORTICULTURAL UNIVERSITIES			
1	Andhra Pradesh Horticultural University	Tadepalli-gudem	Andhra Pradesh
2	University of Horticultural Sciences	Bagalkot	Karnataka
3	Dr Yashwant Singh Parmar Univ of Horticulture and Forestry	Solan	Himachal Pradesh
4	Dr YSR Horticultural University	Veketara-manna-gudem	Andhra Pradesh
5	Uttarakhand University of Horticulture and Forestry	Bharsar	Uttrakhand

Appendix 2. List of ICAR Research Institutes in India

Sl.No.	*Name of Research Institute*	*Location*	*State*
A	**CROP SCIENCES**		
1	ICAR-Central Institute for Cotton Research	Nagpur	Maharashtra
2	ICAR- Central Institute for Jute and Allied Fibres	Kolkata	West Bengal
3	ICAR-National Rice Research Institute	Cuttack	Odisha
4	ICAR -Central Tobacco Research Institute	Rajamundry	Andhra Pradesh
5	ICAR- Directorate of Groundnut Research	Junagarh	Gujarat
6	ICAR-Indian Institute of Oilseeds Research	Hydereabad	Telangana
7	ICAR- Directorate of Rapeseed-Mustard Research	Bharatpur	Rajasthan
8	ICAR-Indian Institute of Rice Research	Hyderabad	Talangana
9	ICAR-Indian Institute of Seed Science	Mau	Uttar Pradesh
10	ICAR-Indian Institute of Millets Research	Hyderabad	Talangana
11	ICAR-Indian Institute of Soybean Research	Indore	Madhya Pradesh
12	ICAR-Indian Agriculture Research Institute	Pusa	New Delhi
13	ICAR-Indian Agriculture Research Institute	Hazaribagh	Jharkhand
14	ICAR-Indian Grassland and Fodder Research Institute	Jhansi	Uttar Pradesh
15	ICAR-Indian Institute of Maize Research	Ludhiana	Punjab
16	ICAR-Indian Institute of Agricultural Biotechnology	Ranchi	Bihar
17	ICAR-Indian Institute of Pulses Research	Kanpur	Uttar Pradesh
18	ICAR-Indian Institute of Sugarcane Research	Lucknow	Uttar Pradesh
19	ICAR-Indian Institute of Wheat and Barley Research	Karnal	Haryana
20	ICAR-National Bureau of Agricultural Insect Resources	Bangaluru	Karnataka

Sl.No.	Name of Research Institute	Location	State
21	ICAR-National Bureau of Agriculturally Important Micro-organism	Mau	Uttar Pradesh
22	ICAR-National Bureau of Plant Genetic Resources	Pusa	New Delhi
23	ICAR-National Centre for Integrated Pest Management	Pusa	New Delhi
24	ICAR- National Institute of Biotic Stress Management	Raipur	Chhatishgarh
25	ICAR-National Centre on Plant Biotechnology	Pusa	New Delhi
26	ICAR-Sugarcane Breeding Institute	Coimbatore	Tamil Nadu
27	ICAR-Vivekanand Parvatiya Krishi Anusandhan Sansthan	Almora	Uttrakhand
B	**HORTICULTURAL SCIENCES**		
28	ICAR-Central Institute for Arid Horticulture	Bikaner	Rajasthan
29	ICAR-Central Institute for Subtropical Horticulture	Lucknow	Uttar Pradesh
30	ICAR-Central Institute for Temperate Horticulture	Srinagar	Jammu and Kashmir
31	ICAR- Central Island Agricultural Research Institute	Portblair	Andman and Nicobar
32	ICAR-Central Plantation Crops Research Institute	Kasargod	Kerala
33	ICAR-Central Potato Research Institute	Shimla	Himachal Pradesh
34	ICAR-Central Tuber Crops Research Institute	Trivendrum	Kerala
35	ICAR-Directorate of Cashew Research	Puttur	Karnataka
36	ICAR-Directorate of Floriculture Research	Pusa	New Delhi
37	ICAR-Directorate of Medicinal And Aromatic Plants Research	Anand	Gujarat
38	ICAR-Directorate of Mashroom Research	Solan	Himachal Pradesh
39	ICAR-Directorate of Onion and Garlic Research	Pune	Maharashtra
40	ICAR-Indian Institute of Horticulture Research	Bangalore	Karnataka

Sl.No.	Name of Research Institute	Location	State
41	ICAR-Indian Institute of Oil Palm Research	Pedavegi	Andhra Pradesh
42	ICAR-Indian Institute of Spices Research	Calicut	Kerala
43	ICAR-Indian Institute of Vegetable Research	Varanasi	Uttar Pradesh
44	ICAR-National Research Centre for Banana	Tiruchirapalli	Kerala
45	ICAR-Central Citrus Research Institute	Nagpur	Maharashtra
46	ICAR-National Research Centre for Grapes	Pune	Maharashtra
47	ICAR-National Research Centre for Litchi	Muzaffarpur	Bihar
48	ICAR-National Research Centre for Orchids	Gangtok	Sikkim
49	ICAR-National Research Centre on Pomegranate	Solapur	Maharashtra
C	NATURAL RESOURCE MANAGEMENT		
50	ICAR-National Research Centre for Seed Spices	Ajmer	Rajasthan
51	ICAR-Central Agro-forestry Research Institute	Jhansi	Uttar Pradesh
52	ICAR-Central Arid Zone Research Institute	Jodhpur	Rajasthan
53	ICAR-Central Research Institute for Dry land Agriculture	Hyderabad	Andhra Pradesh
54	ICAR-Indian Institute of Soil and Water Conservation	Dehradun	Uttrakhand
55	ICAR-Central Soil Salinity Research Institute	Karnal	Haryana
56	ICAR-Indian Institute of Water Management	Bhubaneswar	Odisha
57	ICAR-Directorate of Weed Research	Jabalpur	Madhya Pradesh
58	ICAR Research Complex for NEH Region	Shillong	Meghalaya
59	ICAR Research Complex for Eastern Region	Patna	Bihar
60	ICAR-Indian Institute of Soil Science	Bhopal	Madhya Pradesh

Sl.No.	Name of Research Institute	Location	State
61	ICAR-Indian Institute of Farming System Research	Meerut	Uttar Pradesh
62	ICAR-Indian Institute of Soil Science	Bhopal	Madhya Pradesh
63	ICAR-National Bureau of Soil Survey and Land Use Planning	Nagpur	Maharashtra
64	ICAR-National Institute of Abiotic Research management	Baramati	Maharashtra
65	ICAR- National Organic Farming Research	Gangtok	Sikkim
66	ICAR-National Research Centre for Integrated Farming	Motihari	Bihar
D	**AGRICULTURAL ENGINEERING**		
67	ICAR-Central Institute for Research on Cotton Technology	Mumbai	Maharashtra
68	ICAR-Central Institute of Agricultural Engineering	Bhopal	Madhya Pradesh
69	ICAR-Central Institute Post Harvest Engineering and Technology	Ludhiana	Punjab
70	ICAR-Indian Institute of Natural Resigns and Gums	Ranchi	Jharkhand
71	ICAR-National Institute of Research on Jute and Allied Fibres Technology	Kolkata	West Bengal
E	**AGRICULTURAL EDUCATION**		
72	ICAR-Central Institute for Women in Agriculture	Bhubaneswar	Odisha
73	ICAR-Indian Agricultural Statistics Research Institute	Pusa Campus	New Delhi
74	ICAR-National Academy of Agricultural Research Management	Hyderabad	Telangana
75	ICAR-National Institute of Agricultural Economics and Policy Research	Pusa Campus	New Delhi
F	**AGRICULTURAL EXTENSION**		
76	ICAR-Directorate of Knowledge Management in Agriculture	Pusa Campus	New Delhi
G	**ANIMAL RESEARCH INSTITUTES**		
77	ICAR-Central Avian Research Institute	Bareilly	Uttar Pradesh
78	ICAR-Central Institute for Research on Buffaloes	Hisar	Haryana

Sl.No.	Name of Research Institute	Location	State
79	ICAR-Central Institute for Research on Cattle	Meerut	Uttar Pradesh
80	ICAR-Central Institute for Research on Goats	Mathura	Uttar Pradesh
81	ICAR-Central Sheep and Wool Research Institute	Avikanagar	Rajasthan
82	ICAR-Directorate of Poultry Research	Hyderabad	Telangana
83	ICAR-Indian Veterinary Research Institute	Bareilly	Uttar Pradesh
84	ICAR-National Bureau of Animal Genetic Resources	Karnal	Haryana
85	ICAR-National Dairy Research Institute	Karnal	Haryana
86	ICAR-National Institute of Animal Nutrition and Physiology	Karnal	Haryana
87	ICAR-National Institute of High Security Animal Diseases	Bhopal	Madhya Pradesh
88	ICAR-National Institute of Veterinary Epidemiology and Disease Informatics	Bangalore	Karnataka
89	ICAR-National Research Centre for Equines	Hisar	Haryana
90	ICAR-National Research Centre for Camel	Bikaner	Rajasthan
91	ICAR-National Research Centre on Meat	Hyderabad	Telangana
92	ICAR-National Research Centre on Mithun	Jharnapani	Nagaland
93	ICAR-National Research Centre on Pig	Guwahati	Assam
94	ICAR-National Research Centre on yak	Dirang	Arunachal Pradesh
95	ICAR-Project Directorate on Foot and Mouth Disease	Nainital	Uttrakhand
H	**FISHERIES INSTITUTES**		
96	ICAR-Central Island Fisheries Research Institute	Barrackpur	West Bengal
97	ICAR-Central Institute of Brackish water Aquaculture	Chennai	Tamil Nadu
98	ICAR-Central Institute Fisheries Education	Mumbai	Maharashtra
99	ICAR-Central Institute Fisheries Technology	Cochin	Kerala

Sl.No.	Name of Research Institute	Location	State
100	ICAR-Central Institute of Fresh Water Aquaculture	Bhubaneswar	Odisha
101	ICAR-Central Marine Fisheries Research Institute	Kochi	Kerala
102	ICAR-Directorate of Cold Water Fisheries Research	Bhimtal	Uttarakhand
103	ICAR-National Bureau of Fish Genetic Resources	Lucknow	Uttar Pradesh

Appendix 3: List of International Agricultural Research Institutes/Centres

Sl. No.	*Name of Research Institute/Centre*	*Founded in*	*Location*	*Country*
1.	International Rice Research Institute [IRRI]	1960	Manila	Philippines
2.	International Wheat and Maize Improvement Centre [CIMMYT]	1963	Mexico City	Mexico
3.	International Centre for Tropical Agriculture [CIAT]	1967	Cali	Colombia
4.	International Institute for tropical Agriculture [IITA]	1967	Ibadan	Nigeria
5.	International Potato Centre [CIP]	1971	Lima	Peru
6.	International Crop Research Institute for Semiarid Tropics [ICRISAT]	1972	Hyderabad	India
7.	International Centre for Agriculture Research in Drylans Areas [ICARDA]	1977	Aleppo	Syria
8.	West Africa Rice Development Association [WARDA]	1970	Bouake	Cote divoire
9.	International Plant Genetic Resources Institute [IPGRI]	1974	Rome	Italy
10.	Centre for International Forestry Research [CIFOR]	1992	Jakarta	Indonesia
11.	International Centre for Research in Agroforestry [ICRAF]	1991	Nairobi	Kenya
12.	International Centre for Living Aquatic Resources Management [ICLARM]	1977	Makati City	Philippines
13.	International Food Policy Research Institute [IFPRI]	1975	Washington DC	USA
14.	International Irrigation Management Institute [IIMI]	1984	Colombo	Sri Lanka
15.	International Livestock Research Institute [ILRI]	1995	Nairobi	Kenya
16.	International Service for National Agricultural Research [ISNAR]	1979	Hague	Netherlands

Sl. No.	*Name of Research Institute/Centre*	*Founded in*	*Location*	*Country*
17.	Asian Vegetable Research and Development Centre [AVRDC]	1971	Taiwan City	Taiwan [China]
18.	International Centre for Genetic Engineering and Biotechnology [ICGEB]	1987	Trieste	Italy

Glossary

Chapter 1: Molecular Techniques

Allele specific oligonucleotide (ASO): A technique that allows detection of single base mutations without the need for PCR or gel electrophoresis.

DNA fingerprinting: A technique that is used to distinguish between individuals of the same species using only samples of their DNA; also called genetic fingerprinting, DNA testing, DNA typing, or DNA profiling.

DNA Micro Arrays: A collection of spots attached to a solid support such as a microscope slide where each spot contains one or more single-stranded DNA.

DNA Sequencing: The process of determining the order of the nucleotide bases along a DNA strand.

Expression Cloning: One of the most basic techniques of molecular biology to study protein function.

Fluorescence *in situ* hybridization (FISH): A newer method of analyzing chromosomes, using fluorescent molecules, called **dyes**, to "paint" genes on a chromosome.

Gel electrophoresis: One of the principal tools of molecular biology. The basic principle is that DNA, RNA, and proteins can all be separated by means of an electric field.

Gene Cloning: Gene cloning is the process of making several copies of a gene.

Molecular Genetics: Study of structure and function of genes at the molecular level.

Northern Blot Technique: A method of detecting specific type of RNA molecule among of a set of different samples of RNA.

Polymerase Chain Reaction [PCR]: An extremely versatile technique for copying DNA.

Southern Blot Technique: A technique of detecting a small number of DNA fragments which may be present in a complex mixture.

Western Blot Technique: A technique of detecting specific protein in a sample.

Chapter 2: DNA Isolation

DNA Isolation: The process of isolating [separating] DNA from plant and animal tissues.

DNA Probe: A small nucleotide sequence [15-30bp] of DNA or RNA which is used to detect the presence of complimentary sequences of nucleic acid in a sample.

DNA: Deoxyribose nucleic acid; the genetic material.

Gene Detection: The process of detecting the presence of a particular gene in the DNA sample.

Northern Blotting: A technique used in molecular biology research to determine the presence of a specific mRNA from a set of different samples of RNA.

PCR: A technique widely used in molecular biology for gene amplification and DNA polymorphism.

Southern blotting: A method routinely used in molecular biology to detect the presence of a specific DNA sequence (a gene) in a DNA sample.

Western Blotting: A method of detecting the presence of a specific proteins in a given sample of tissues or extract using antibody specific to that protein.

Chapter 3: Gene Cloning

Gene cloning: The process by which multiple copies of a specific gene (DNA) can be made after isolation of such gene or DNA; also known as DNA cloning or genetic cloning or molecular cloning or gene amplification.

Gene Library: The collection of gene clones is called a DNA library or gene library.

Chapter 4: Polymerase Chain Reaction

Annealing: The process of hybridization of a mixture of DNAs in single stranded form and detection of homologous regions in the DNA of two species involved.

Denaturation: The separation of two strands of DNA in to single strands.

Polymerase chain reaction (PCR): A technique which is widely used in molecular biology for gene amplification, DNA polymorphism, gene detection, *etc.*

Probes: Complimentary sequences of nucleotide bases to the specific mRNA sequence of interest.

Renaturation: Union of denatured strands of DNA on cooling.

Chapter 5: Gene Detection Techniques

DNA Isolation: The process of isolating [separating] DNA from plant and animal tissues.

DNA Probe: A small nucleotide sequence [15-30bp] of DNA or RNA which is used to detect the presence of complimentary sequences of nucleic acid in a sample.

DNA: Deoxyribose nucleic acid; the genetic material.

Gene Detection: The process of detecting the presence of a particular gene in the DNA sample.

Northern Blotting: A technique used in molecular biology research to determine the presence of a specific mRNA from a set of different samples of RNA.

PCR: A technique widely used in molecular biology for gene amplification and DNA polymorphism.

Southern blotting: A method routinely used in molecular biology to detect the presence of a specific DNA sequence (a gene) in a DNA sample.

Western Blotting: A method of detecting the presence of a specific proteins in a given sample of tissues or extract using antibody specific to that protein.

Chapter 6: Genotyping and Phenotyping

Genotype: Genetic constitution of an individual.

Genotyping: Molecular characterization of an individual.

Phenotype: External appearance of an individual.

Phenotyping: Morphological characterization of an individual.

Chapter 7: QTL Mapping

Composite Interval mapping: A technique of QTL mapping in which markers are used as co-factors.

Gene mapping: A molecular technique which is used to identify genes responsible for expression of specific traits.

Interval Mapping: A technique of QTL mapping in which two markers are used.

Multiple Interval mapping: A technique of QTL mapping in which several markers are used.

QTL mapping: The technique of locating genes that control expression of quantitative traits.

QTL: A region of DNA which is associated with the expression of a quantitative trait.

Single marker Analysis: A technique of QTL mapping in which one marker is used.

Chapter 8: DNA Fingerprinting

Circular DNA: DNA that has a ring or circular shape. Such DNA is usually found in prokaryotes, chloroplasts and mitochondria.

Cytoplasmic DNA: DNA which is found in the cytoplasm in chloroplasts and mitochondria.

Deoxyribose: A pentose sugar having five carbon atoms.

DNA fingerprinting: A technique by which an individual can be identified at molecular level.

DNA: A nucleic acid that carries the genetic information in the cell and is capable of self-replication and RNA synthesis.

Double Stranded DNA: DNA having spirally arranged two strands.

Left handed DNA: DNA with coiling of the helix in the left direction such as Z-DNA.

Linear DNA: DNA that has thread like structure with both ends free. Such DNA is found in eukaryotes.

Nuclear DNA: DNA which is found in the nucleus in the chromosomes.

Repetitive DNA: DNA in which small sequences of bases are repeated several hundred times. Such DNA is found in all eukaryotes except yeast.

Right handed DNA: DNA with coiling of the helix in the right direction, a usual form of DNA.

Single Stranded DNA: DNA having one helix only. Found in some bacteriophage.

Unique DNA: The DNA segment having single copy per genome.

Chapter 9: Methods of Foreign Gene Transfer

Direct Gene Transfer: Introduction of foreign DNA into plant cells without the involvement of biological agents such as Agrobacterium.

DNA Delivary via growing pollen tubes: In this method, stigmas are cut after pollination and the DNA solution is applied onto the cut surface for easy diffusion through the germinating pollen tube into the ovule.

Fibre Mediated transformation: Delivery of foreign DNA in to the cell cytoplasm and nucleus by silicon carbide fibres.

Gene Gun Method: Delivery of foreign DNA into host plant cells using high velocity metal particles through gene gun.

Genetic transformation: The process of transfer, integration and expression of transgene in the host cells.

Indirect Gene Transfer: The vector mediated gene transfer or biological method of gene transfer or Agrobacterium mediated gene transfer.

Laser induced transformation: Delivery of foreign DNA into plant cells with a laser micro-beam.

Lipofection: Introduction of DNA into cells via liposomes.

Macro-injection: Delivery of foreign DNA in to host cells by injection of plasmid DNA into the lumen of developing inflorescence using hypodermic syringe.

Microinjection: Delivery of foreign DNA in to host cells by microscopic needles.

Chapter 10: Molecular Markers

Biochemical Marker: Markers that are related to variation in proteins and amino acid banding pattern.

Cytological Marker: Markers that are related to variation in chromosome number, shape and size.

DNA Marker: A gene or other fragment of DNA whose location in the genome is known. A genetic marker is a known DNA sequence.

Epistasis: The interaction between genes. Epistasis takes place when the action of one gene is modified by one or several other genes.

Genotyping: The process of determining the genotype of an individual with a biological assay

Locus (plural loci): A fixed position on a chromosome, such as the position of a gene or a biomarker (genetic marker).

Marker assisted selection: Selection for specific alleles (which affect a trait of interest) using genetic markers.

Marker: Any genetic element (locus, allele, DNA sequence or chromosome feature) which can be readily detected by phenotype, cytological or molecular techniques, and used to follow a chromosome or chromosomal segment during genetic analysis.

Molecular Breeding: Improvement of crop plants for various economic characters through indirect selection for linked molecular markers.

Morphological Marker: In plant breeding, markers that are related to variation in shape, size, colour and surface of various plant parts.

Phenotype: Any observable characteristic of an organism, such as its morphology, development, biochemical or physiological properties, or behavior.

Pleiotropy: Influence of a single gene on multiple phenotypic traits.

Polymorphism: Presence of multiple alleles of a gene within a population, usually expressing different phenotypes

Chapter 11: Marker Assisted Selection

Biochemical Marker: Markers that are related to variation in proteins and amino acid banding pattern.

Cytological Marker: Markers that are related to variation in chromosome number, shape and size.

DNA Marker: A gene or other fragment of DNA whose location in the genome is known. A genetic marker is a known DNA sequence.

Epistasis: The interaction between genes. Epistasis takes place when the action of one gene is modified by one or several other genes.

Genotyping: The process of determining the genotype of an individual with a biological assay

Locus (plural loci): A fixed position on a chromosome, such as the position of a gene or a biomarker (genetic marker).

Marker assisted selection: Selection for specific alleles (which affect a trait of interest) using genetic markers.

Marker: Any genetic element (locus, allele, DNA sequence or chromosome feature) which can be readily detected by phenotype, cytological or molecular techniques, and used to follow a chromosome or chromosomal segment during genetic analysis.

Molecular Breeding: Improvement of crop plants for various economic characters through indirect selection for linked molecular markers.

Morphological Marker: In plant breeding, markers that are related to variation in shape, size, colour and surface of various plant parts.

Phenotype: Any observable characteristic of an organism, such as its morphology, development, biochemical or physiological properties, or behavior.

Pleiotropy: Influence of a single gene on multiple phenotypic traits.

Polymorphism: Presence of multiple alleles of a gene within a population, usually expressing different phenotypes

Chapter 12: Transgenic Breeding

Biotechnology: Application of various biological organisms or processes for mass production of useful substances or products for industry, medicine and agriculture.

Gene Revolution: Quantum jump in the productivity of various field crops through the use of transgenic cultivars.

Protoplasts: Naked cells or cells without cell wall.

Transgenes: Foreign genes or modified genes of the same species which are used for development of transgenic individuals.

Transgenic Breeding: Genetic improvement of crop plants, domestic animals and useful micro-organisms, through application of biotechnology, in relation to their economic use for mankind.

Transgenic: Individuals containing altered or transgenes from an unrelated organism. It involves taking genes from one species and inserting them into another species to get that trait expressed in the offspring.

Chapter 13: Transgenic Male Sterility

Antisense: Short DNA or RNA sequences, termed oligonucleotides, which are complementary to a specific gene sequence.

Male Sterility: A condition in which either pollen is absent or non-functional in flowering plants.

Restorer capacity: The ability and efficiency of a restorer line to restore male fertility efficiently in a wide variety of male sterile lines.

Transgenic Male Sterility: The male sterility which is induced by the technique of genetic engineering.

Chapter 14: Mapping Populations

Back Cross: Crossing of F_1 with either of its parents.

Doubled Haploid: An individual obtained by chromosome doubling of a haploid individual.

Gene Mapping: The process of locating genes on specific position in a chromosome.

Haploid: An individual with basic set of chromosomes

Isogenic Line: Lines with single gene difference.

Mapping Populations: Various type of plant materials which are used for gene mapping.

NILs: Near isogenic lines.

Pure Line: A line obtained by selfing of a self pollinated plant.

RILs: Recombinant inbred lines.

Test Cross: Crossing of F_1 with its homozygous recessive parent.

Chapter 15: Introduction to Genomics

Functional Genomics: The study of function of all genes present in the entire genome of a living organism.

Functional Proteomics: The study of functions of all proteins found in a living organism.

Genome: The basic set of chromosomes. In a genome, each type of chromosome is represented only once.

Genomics: The study of structure and function of entire genome of a living organism.

Metabolomics: The study of all metabolic pathways in a living organism.

Proteome: A complete set of proteins encoded by a genome.

Proteomics: The study of structure and function of all proteins in an individual.

Structural Genomics: The study of the structure of entire genome of a living organism.

Structural Proteomics: The study of structure of all proteins found in a living organism.

Chapter 16: Introduction to Bioinformatics

Bioinformatics: The computer aided study of biology particularly genetics and molecular biology.

Candidate variety: A variety to be registered under Plant Variety Protection Act is referred to as candidate variety.

Example Variety: A variety that is used for comparison for a particular character is called example variety.

Extant Variety: All released, notified and unprotected varieties.

Farmers' Variety: A variety that has been developed by a farmer and used for commercial cultivation for several years is called farmers variety.

Reference Variety: All released and notified extant varieties of common knowledge which are in seed production chain.

Chapter 17 and 18: Gene Revolution and Green Revolution

Gene Revolution: Quantum jump in the productivity of various field crops through use of transgenic cultivars.

Green Revolution: Green revolution refers to the quantum jump in the productivity of cereals especially wheat and rice through the use of dwarfing genes.

Green Revolution: Quantum jump in the productivity of cereals through use of dwarfing genes.

Transgenic Breeding: Genetic improvement of crop plants through the application of agricultural biotechnology.

Transgenic Plants: Genetically engineered plants.

Chapter 19: Nanotechnology in Crop Improvement

Gene Delivery: Successful transfer of a gene into plant cell.

Nano-machines: Machines smaller than 100 nanometers.

Nano-particles: Particles smaller than 100 nanometers.

Nano-technology: A branch of engineering that deals with things smaller than 100 nanometers.

Chapter 20: Reverse Breeding

Forward Breeding: Breeding approaches which promote genetic recombination and reduce recovery of parental types.

Reverse Breeding: Breeding approaches which suppress genetic recombination and promote recovery of parental types.

Chapter 21: Space and Speed Breeding

Recoverable satellite: The satellite which returns back to earth after moving in the space.

Space breeding: The technique of sending seeds or microorganisms into space (200-400km) on a recoverable satellite or in other recoverable spacecraft.

Speed Breeding: Speed breeding technology is an indoor plant growing technique. It shortens harvest time (2-3x faster) of crops like wheat and barley.

Some Useful References

Humphreys, M. 2005. Molecular Breeding for Genetic Improvement of Forage Crops and Turf. Wagenigen Press.

Jannink J. L., Lorenz A. J., and Iwata H.(2010. Genomic selection in plant breeding: from theory to practice. Brief Functional Genomics 9:166-177

Lesk, A. M. 2002. Introduction to Bio-informatics. OUP, Oxford.

Newbury, H. J. (ed.) 2003. Plant Molecular Breeding. The Uni. Birmingham, U.K.

Prim Rose, S. B. 2006. Principles of Gene manipulation and Genomics. Blackwell Publishing Inc.

Ribaut, J. M. and Hoisington, D.A. 1998. Marker Assisted Selection: New Tools and Strategies. Trends Plant Sci., 3236-239.

Sensen, C. W. (ed.) 2002. Essentials of Genomics and Bio-informatics. Wiley-VCH, GmbH, Weinheim.

Sevan Bode (ed.) 2013. Molecular Markers and Marker-Assisted Breeding in Plants. Plant Science Department, South Dakota State University, Brookings, USA.

Singh Phundan 2014. Plant Breeding: Molecular and New Approaches. 3rd Edition, Kalyani Publishers, NNew Delhi.

Singh, R. K. 2010. Molecular Plant Breeding: Principle, Method and Application, Studium Press.

Stephen P. Moose and Rita H. Mumm (2008) Molecular Plant Breeding as the Foundation for 21st Century Crop Improvement, Plant Physiology 147:969-977

Voosen P. 2009. Molecular Breeding Makes Crops Hardier and More Nutritious Markers, knockouts and other technical advances improve breeding without modifying genes, Scientific American

Young, N. D. 1999. A Cautiously Optimistic Vision for Marker Assisted Breeding. Mol Breed. 5:505-510.

Yunbi Xu 2010. Molecular Plant Breeding. International Maize and Wheat Improvement Centre (CIMMYT), China.

Yunbi Xu 2012. Molecular Plant Breeding. CABI Publishing.

Key References

A. Molecular Biology

1. Strickberger, M. W. 1985. Genetics. 3rd ed. Macmillan Publishing Co., New York, USA.
2. Watson, J. D. 1976. Molecular Biology of Gene3rd edition. Benjamin, Menlo Park, California, USA.
3. Friefelder, D. 1987. Molecular Biology 2nd ed. Jones Bartlett Publishers, Inc. USA.
4. Phundan Singh 2015. Molecular Plant Breeding. Kalyani Publishers, New Delhi.
5. Phundan Singh 2018. Plant Breeding: Molecular and New Approaches. Fourth Edition. Kalyani Publishers, New Delhi.
6. Phundan Singh 2012. Molecular Genetics. Kalyani Publishers, New Delhi.
7. Phundan Singh 2012. Molecular Genetics: At A Glance. Kalyani Publishers, New Delhi.
8. Phundan Singh 2012. Objective Molecular Genetics. Kalyani Publishers, New Delhi.

B. Plant Biotechnology

9. Bajaj, Y.P.S. 1994. Somatic Hybridization in Crop Improvement. Springer Verlag.
10. Bhojwani, S. S. (ed.) 1990. Plant Tissue Culture: Applications and Limitations. Elsevier, Amesterdam.
11. Bhojwani, S. S. and Bhatnagar, S. P. 1990. Embryology of Angiosperms. Vikas Publishing Co., New Delhi.
12. Bhojwani, S. S. and Rajdan, M. K. 1996. Plant Tissue Culture: Theory and Practice. Revised Edition. Elsevier, Amesterdam.

13. Chawla, H. S. 2002. Introduction to Plant Biotechnology. 2nd ed. Oxford and IBH Publishing Co.PVT. LTD., New Delhi.
14. Khanna, V. K. 1999. Plant Tissue Culture. Kalyani Publishers, New Delhi.
15. Phundan Singh 2013. Introduction to Biotechnology. Second edition. Kalyani Publishers, New Delhi.
16. Phundan Singh 2019. Plant Biotechnology. Second Edition. Kalyani Publishers, New Delhi.
17. Phundan Singh 2013. Principles of Plant Biotechnology. Kalyani Publishers, New Delhi.
18. Phundan Singh 2016. Objective Plant Biotechnology. Kalyani Publishers, New Delhi.
19. Phundan Singh 2016. Plant Biotechnology At A Glance. Kalyani Publishers, New Delhi.
20. Prakash, J and Pierik, R.L.M. (eds). 1993. Plant Biotechnology: Commercial Prospects and Problems. Oxford and IBH Publishing Co.PVT. LTD., New Delhi.
21. Rajdan, M. K. 1993. An Introduction to Plant Tissue Culture. Oxford and IBH Publishing Co.PVT. LTD., New Delhi.
22. Sensen, C. W. (ed) 2002. Essentials of Genomics and Bioinformatics. Wiley-VCH.
23. Singh, B. D. 2006. Plant Biotechnology. Kalyani Publishers, New Delhi.
24. Vasil, I. K. and Thorpe T. A. 1994. Plant Cell and Tissue Culture. Kluwer Academic Publishers, Dordrecht, Netherlands.
25. Zaitlin, M., Day, P. and Hollaender, A. 1985. Biotechnology in Plant Science: Relevance to Agriculture in the Eightees. Academic Press, N.Y.

Books by Dr. Phundan Singh

A. Plant Breeding

1. Essentials of Plant Breeding
2. Fundamentals of Plant Breeding
3. Plant Breeding: Molecular and New Approaches
4. Molecular Plant Breeding
5. Commercial Plant Breeding
6. Commercial Plant Breeding: At A Glance
7. Commercial Plant Breeding: Objective
8. Objective Science of Plant Breeding
9. Genetic Principles and Plant Breeding
10. Genetics and Plant Breeding: Comparative Analysis
11. Principles of Plant Breeding[For PG Students]
12. Plant Breeding: At a Glance
13. Plant Breeding [For Under Graduate Students]
14. Principles of Plant Breeding [For UG Students]
15. Practical in Crop Breeding
16. Practical and Numerical in Plant Breeding
17. Numerical Problems in Plant Breeding and Genetics
18. Glossary of Plant Breeding and Genetics
19. Breeding Crop Plants for Stress Resistance
20. Introduction to Resistance Plant Breeding

21. Resistance Plant Breeding: Objective
22. Resistance Plant Breeding: At A Glance
23. Plant Breeding: Related Legislations
24. Concepts in Plant Breeding
25. Principles of Seed Technology
26. Seed Technology: At a Glance
27. Objective Seed Technology

B. Cotton Breeding and Genetics

28. Cotton Breeding
29. Heterosis Breeding in Cotton
30. Breeding Hybrid Cotton
31. Breeding Transgenic Bt. Cotton
32. Basics of Bollgard Cotton
33. Basics of Bollgard Cotton: At A Glance
34. Basics of Bollgard Cotton: Objective
35. Cotton Improvement in India
36. Sankar Kapas [In Hindi]
37. Glimpses of Cotton Breeding
38. Cotton Genetics
39. Cotton

C. Genetics

40. Elements of Genetics
41. Genetics
42. Principles of Genetics [For UG Students]
43. Fundamentals of Genetics
44. Plant Genetics
45. Objective Genetics
46. Genetics: At a Glance
47. Objective Genetics and Plant Breeding
48. Objective Genetics and Plant Breeding [Hindi Edition]
49. Glossary cum Dictionary of Genetics and Plant Breeding
50. Molecular Genetics [Objective]
51. Molecular genetics [At a Glance]
52. Molecular Genetics [Subjective]

53. Genetics and Man
54. Genetics Today

D. Quantitative Genetics

55. Biometrical Techniques in Plant Breeding
56. Application of Biometrical Techniques in Plant Breeding [Hindi Edition]
57. Objective Quantitative Genetics
58. Quantitative Genetics: At a Glance
59. Quantitative Genetics
60. Introduction to Agricultural Statistics
61. Agricultural Statistics: Objective
62. Agricultural Statistics: At A Glance

E. Plant Biotechnology

63. Introduction to Biotechnology
64. Plant Biotechnology
65. Principles of Plant Biotechnology
66. Plant Biotechnology: At a Glance
67. Objective Plant Biotechnology
68. Fundamentals of Molecular Biology and Plant Biotechnology
69. Objective Molecular Biology and Plant Biotechnology
70. Molecular Biology and Plant Biotechnology: At A Glance

F. Intellectual Property Rights

71. IPR and Plant Breeders' Rights [Subjective]
72. IPR and Plant Breeders' Rights [Objective]
73. IPR and Plant Breeders' Rights [At a Glance]
74. Introduction to Intellectual Property Rights
75. Intellectual Property Rights: At A Glance
76. Intellectual Property Rights: Objective

G. Miscellaneous Books

77. Elements of Baby Corn
78. Baby Corn Farming for Higher Income
79. A Guide for Competitive Examinations of Agriculture
80. Objective General Agriculture for Competitive Examinations
81. Autobiography of Phundan Singh

82. Concepts in Plant Breeding [Objective Analysis]
83. Concepts in Plant Breeding [Question and Answer]
84. Principles of Plant Breeding [Objective Analysis]
85. Principles of Plant Breeding [Question and Answer]
86. Principles of Crop Improvement [Objective Analysis]
87. Principles of Crop Improvement [Question and Answer]

Printed at Repro India Ltd.